The Analogue
Alternative

Studies in the History of Science, Technology and Medicine
Edited by John Krige, CRHST, Paris, France

Studies in the History of Science, Technology and Medicine aims to stimulate research in the field, concentrating on the twentieth century. It seeks to contribute to our understanding of science, technology and medicine as they are embedded in society, exploring the links between the subjects on the one hand and the cultural, economic, political and institutional contexts of their genesis and development on the other. Within this framework, and while not favouring any particular methodological approach, the series welcomes studies which examine relations between science, technology, medicine and society in new ways e.g. the social construction of technologies, large technical systems.

Other Titles in the Series

Volume 1
Technological Change: Methods and Themes in the History of Technology
Edited by Robert Fox

Volume 2
Technology Transfer out of Germany after 1945
Edited by Matthias Judt and Burghard Ciesla

Volume 3
Entomology, Ecology and Agriculture: The Making of Scientific Careers in North America, 1885–1985
Paolo Palladino

Volume 4
The Historiography of Contemporary Science and Technology
Edited by Thomas Söderquist

Volume 5
Science and Spectacle: The Work of Jodrell Bank in Post-war British Culture
Jon Agar

Volume 6
Molecularizing Biology and Medicine: New Practices and Alliances, 1910s-1970s
Edited by Soraya de Chadarevian and Harmke Kamminga

Volume 7
Cold War, Hot Science: Applied Research in Britain's Defence Laboratories 1945-1990
Edited by Robert Bud and Philip Gammett

Volume 8
Planning Armageddon: Britain, the United States and the Command of Western Nuclear Forces 1945–1964
Stephen Twigge and Len Scott

Volume 9
Cultures of Control
Edited by Miriam R. Levin

Volume 10
Science, Cold War and the American State: Lloyd V. Berkner and the Balance of Professional Ideals
Alan A. Needell

To my parents Catherine and James, and to Marjan

The Analogue Alternative

THE ELECTRONIC ANALOGUE COMPUTER IN BRITAIN AND THE USA, 1930–1975

James S. Small
formerly of
The Universities of Manchester, England, and Maastricht
The Netherlands

Routledge
Taylor & Francis Group

LONDON AND NEW YORK

First published 2001 by Routledge
4 Park Square, Milton Park, Abingdon, Oxon OX14 4RN
605 Third Avenue, New York, NY 10017

First issued in paperback 2013

*Routledge is an imprint of the Taylor & Francis Group,
an informa business*

Copyright © 2001 by Taylor & Francis

Typeset by Expo Holdings, Malaysia

British Library cataloguing in Publication Data
A catalogue record for this book is available from the British
Library

ISBN 13: 978-0-415-27119-6 (hbk)
ISBN 13: 978-0-415-86299-8 (pbk)

Contents

List of Figures

List of Tables

Acknowledgments

The illustrations that appear in this book of the MIT Differential Analyzer and the Rockefeller Differential Analyzer, appear with the permission of the MIT Museum. The illustration of the RCA Laboratories Project Typhoon analogue computer appears courtesy of the David Sarnoff Collection Inc. The remaining illustrations appear courtesy of a private collection. Every effort has been made to contact all the copyright-holders, if any have been inadvertently overlooked they are kindly requested to contact the author.

ACKNOWLEDGEMENTS

I thank David Edgerton for his enthusiasm and guidance during the course of the research for and writing of this book. I further thank Geoffrey Tweedale and Geoffrey Bowker, both formerly of the National Archive for the History of Computing, at the University of Manchester, and John V. Pickstone for their encouragement and assistance. I thank all those I have met in, or through, the Department of Science and Technology Policy and the Centre for the History of Science, Technology and Medicine at the University of Manchester for the help and advice they have given.

This project was, in part, supported by the Economic and Social Research Council in Britain, and in the USA by the Charles Babbage Institute for the History of Information Processing and a summer internship at the IEEE, Center for the History of Electrical Engineering. I gratefully acknowledge the sponsorship of this internship by the IEEE Foundation Life Members Fund. I also wish to thank William Aspray, Arthur Norberg and J. A. N. Lee for the interest they have shown and for the help they have given. Further, thanks go to those who raided their own archives to provide me with documents and photographs, especially Robin Penfold, Mike Brown, Barry Thompson, Keith Knock, Derek Cartwright, Frank Kingdon, Henry Paynter, John McLeod, Walter Karplus, and Granino and Theresa Korn.

I would also like to thank the anonymous referee and John Krige for their helpful comments. Finally, I want to thank Kenneth Gilbert and Marjan Groen, who commented on drafts of this book and helped to shape the final version. Special thanks go to Marjan Groen, my wife, for living with this project as much I have, for her support, and above all for being a great partner to live with – the journey is the destination.

CHAPTER 1

Introduction

In 1968 Albert Hass of the Philadelphia Inquirer wrote:

GIGANTIC COMPUTER GOES TO JUNK PILE

It's hard to become emotionally involved when a machine dies. But the veteran computer scientists at Johnsville Naval Air Development Centre did feel a twinge of nostalgia this week over the death of "Typhoon". In its day, Typhoon was the largest, and most able analogue computer in the world. It made a significant contribution to early Naval missile development and subsequently played a cogent role in the design of scores of aircraft. But by this week the last of its ponderous components had been unceremoniously unplugged and stored and Typhoon had become $3 million worth of electronic junk.[1]

In the past two decades it has become almost inconceivable that an article beginning "Gigantic Computer ..." would subsequently refer to anything other than an electronic digital computer. Or indeed, that there was a time when readers might have stopped to ask themselves whether the article that followed would be discussing electronic analogue or digital computers. The present-day use of the word "computer" refers exclusively to the electronic digital computer. The digital computer is the dominant computing technology of today, but from the late 1940s to the mid-1970s, whether a computer was analogue or digital was not only a pertinent question but a non-trivial one. In a wide range of applications in engineering and to a lesser extent in science, the electronic analogue was an alternative computing technology that was in direct competition with digital computing. For more than three decades from the end of World War II electronic analogue computers co-evolved with developments in electronic devices, feedback amplifier design, control theory, servomechanisms, modelling techniques and simulation, aeronautics, digital computers and computer science.

The general-purpose electronic analogue computers, with which this book is primarily concerned, were fundamentally mathematical machines, used primarily for the solution of differential equations and the simulation of dynamic systems. In engineering and science a great many problems can be defined by differential equations. However, until the advent of high-speed computers, obtaining the solutions to these equations by analytical and mechanical methods was often prohibitively time-consuming. Consequently, many engineering problems either remained unsolved, or approximations—often quite crude ones—were used instead. Thus the high-speed and direct mathematical characteristics of electronic analogue computers made them particularly appropriate and welcome aids to computation in engineering design. For example, electronic analogue computers were used in electrical and mechanical engineering design, in the study of chemical and biological processes and in nuclear reactor design studies.

Yet both in the USA and Britain the principal driving force behind the development of electronic analogue computers was the demand for aids to computation that arose in the design and development of guided missiles and aircraft. Indeed, from the mid-1940s to the mid-1970s, in the USA and Britain aeronautics was the dominant context that sustained their development and use. In the USA the bulk of the postwar work to develop electronic analogue computer systems took place within R&D programmes in industry and academia and was funded directly by the US military. In Britain similar work was undertaken in government-funded research establishments on behalf of the armed forces. Yet military and government interests clearly lay not in the development of analogue computing *per se*, but in the rapid development of new weapon systems and in reduction of the escalating costs of such programmes. In the development of missiles and aircraft, the costs due to field trials on full-scale prototypes were extremely high. Computers offered the possibility of considerable savings if the number of test flights could be minimised by in-laboratory tests and flight simulations. The combination of economic considerations and the nature of the computational demands arising from aeronautics programmes provided a compelling rationale for the development of electronic analogue computers.

In the USA firms began to manufacture general-purpose electronic analogue computers on a commercial basis in the late 1940s, and by the mid-1950s more than a dozen firms were manufacturing computer systems. These ranged in size from desk-top to room-size, and in cost from a few thousand dollars to several hundreds of thousands of dollars. The situation was similar in Britain, but hardware development and commercialisation ran a few years behind that in the USA. By the late 1950s almost every major aircraft and aerospace company in the USA and Britain, as well as every military aircraft and missile development and test facility, had a general-purpose electronic analogue computer system. In Europe too, electronic analogue computer development and use was widespread, especially in France and Germany, as well as in the Soviet Union and Japan. In the USA, analogue and later hybrid computers (which combined analogue and digital technologies) were used in the design and development of ICBMs and were used by NASA and its contractors during America's manned space programmes, including Apollo. In Britain, analogue and hybrid computers were used in the development of a wide range of guided missiles such as the Rapier surface-to-air system and in military and civil aircraft development including Concorde, the first supersonic airliner.

The parallel development of electronic analogue and digital computers was accompanied by the emergence of two distinct communities consisting of designers/developers, manufacturers and users. These communities were generally referred to, and referred to themselves, as the analogue and the digital "men". The boundary between the communities was often rather fuzzy, such that becoming an analogue or a digital "man" was not an irreversible choice and there was overlap, but the fact that a boundary and sense of community existed, was openly acknowledged. The commercialisation and increased use of electronic analogue and digital computers was accompanied by a debate over the relative merits of the two alternative computing technologies and associated techniques. Though technical merits favoured neither outright, proponents of the digital computer invoked a progressive rhetoric, portraying the electronic analogue computer as outdated and as an anachronism soon to be swept aside.

Nevertheless, by the late-1950s many individuals had invested a significant proportion of their professional careers in either designing, developing or becoming expert in the use of electronic analogue computers. For many of these individuals the real appeal of electronic analogue computers went beyond high computing speeds and the capacity for solving differential equations. It lay instead in their ability to incorporate traditional engineering design practices, characterised by graphical methods, scale models, empirical trial-and-error and parameter variation methods, in a new electronic form. By redefining scale model building techniques and enabling dynamic systems to be simulated in real time, they enhanced traditional engineering design practices. Furthermore, electronic analogue computation minimised the mathematical and emphasised the experimental and visual content of the design process. They provided a new design tool that helped engineers bridge the gap between the limits of theory and practice, and the complex real-world systems that they were constructing.

However, this cultural appeal cannot fully explain the development and success of the electronic analogue computer. Indeed, many other factors had a role in shaping the growth of a distinct postwar electronic analogue computer industry, including the nature, increasing size and complexity of problems in engineering design, the corresponding increase in demand for aids to computation, technological and intra-firm competition, Cold War politics, and thus institutional actors such as military agencies and governments. Yet as we shall see, several of these factors also had a role in the decline and disappearance of large-scale general-purpose electronic analogue (and hybrid) computers.

ELECTRONIC ANALOGUE COMPUTERS IN THE HISTORY OF COMPUTING

In 1988 Michael S. Mahoney, Professor of History and History of Science at Princeton University, wrote a critical review entitled *The History of Computing in the History of Technology*. In it he noted "A look at the literature shows that, by and large, historians of computing are addressing few of the questions that historians of technology are now asking." Furthermore that "Reading their accounts makes it difficult to see the alternatives, as the authors themselves lose touch with a time when they did not know what they now know."[2]

Indeed, until the 1980s historians of computing paid little attention to historiographical issues. The bulk of the existing and new literature is participant history, written by pioneers about pioneers. Accounts generally focus on successes, firsts and facts, and though often rich in detail, they adopt a distinctly narrow frame of reference.[3] This results in a loss of historical contingency as well as context. The existence of alternative paths, and doubts as to the future direction of technical change, are eliminated by the joint influences of hindsight and personal involvement. However, since the early 1980s the influence of a small community of professional historians writing on computing has narrowed the gap between "best practice" in the history of technology and the history of computing.[4] Moreover, since the mid-1980s business and economic historians and sociologists of technology have also made valuable and influential contributions to the history of computing literature. Thus, overall there has been a significant increase in the depth and breadth of scholarship in the field during the 1980s.[5]

Nevertheless, examining the history of computing literature, we find little evidence even of the existence of a postwar electronic analogue alternative. The literature is almost exclusively devoted to the digital computer. Moreover, there is an almost total absence of dissent to the orthodoxy of the digital computer "success story". Where analogue computing is included, it is dealt with in one of two ways.

Firstly, there are a few accounts that describe, or include, aspects of electronic analogue computing. Most of these are pioneer histories of inventor and apparatus. However, even the most detailed account does little to scratch the historical surface, either in terms of electronic analogue computing as business history, military enterprise, alternative competing computing technology, technical controversy, or as a product and tool of engineering design.[6]

Secondly, there are several short and longer works that incorporate developments in analogue computing devices into longer general histories of computing that culminate in the modern electronic digital computer.[7] Analogue devices are dealt with up to, but not beyond the advent of postwar stored-programme digital computers. By omitting any further reference to analogue computer development, these accounts suggest that the centuries-long tradition of parallel developments in analogue and digital computing devices ended shortly after the end of World War II, with the future belonging to digital technology. The analogue tradition is generally portrayed as having reached its zenith in the form of the pre-World War II wheel-and-disk differential analyzers. The existence of a postwar analogue tradition is overlooked or in fleeting references underestimated or misrepresented. In both cases the decline of the analogue tradition is rapid, uncontroversial and unequivocal.

So powerful is the digital computer success story that even in the late-1990s this quite untenable treatment of the history of analogue computing within the larger field of the history of computing is still prevalent. Published in 1996, *Computer: A History of the Information Machine*, by Martin Campbell-Kelly and William Aspray, is the most recent victim of the presentist historical perspective.[8] While appearing to give a balanced account of the role of analogue computing in the 20th century, the book repeatedly mis-interprets and mis-represents just that. For example, in chapter 1, Campbell-Kelly and Aspray state, "Also beginning in the nineteenth century and reaching maturity in the 1920s and 1930s was a separate tradition of analogue computing."[9] On the contrary, as we shall see, the analogue computing tradition continued to develop and grow well into the 1970s and 1980s. Indeed, nowhere in their book is the existence of post-World War II electronic-analogue computing mentioned. Worst still, this book makes a number of serious errors. It suggests that only analogue computers work by the concept of analogy. While, in fact, both analogue and digital computers were used to build analogues (or models) of physical systems. The authors also fail to explain the fundamental differences between analogue and digital computing, i.e. that the former operates on continuous data and the latter operates on discrete data quantities. Without this understanding one cannot properly understand how and why electronic analogue and digital computing techniques and technologies were regarded as alternatives.

Taken as a whole the history of computing literature does little to question, and much to re-enforce, the presentist view which portrays the present-day dominance of digital computing technology as a foregone conclusion, and the only possible, rational outcome of the competition between analogue and digital technology.

However, one major exception does exist: Paul Edwards' book, *The Closed World: Computers and the Politics of Discourse in Cold War America*, published in 1996.[10] In his analysis of the development of computers and computing Edwards makes explicit the political dimension, and emphasises the influence of ideologies and popular culture. He employs the concepts of metaphor and discourse to show technological change as the product of complex interactions between scientists, engineers, the military, and governments, and the cultural frames in which these actors are imbedded. For Edwards the history of computing teaches us that technological change consists of technological choices which are tied inextricably to political choice and social values at every level.

Edwards' book is based upon, and provides, an implicit critique of existing computer history, which he criticises, for its reliance on "the tropes of progress and revolution" to structure the narrative.[11] Edwards adopts an analytical framework based on contingency in history, he is thus not constrained to exclude or marginalise alternatives, or indeed portray them as "failures". In his book Edwards has independently made a number of similar points to those made here regarding the analogue to digital "paradigm" shift. Edwards too points up the social inertia to change resulting from the commitment by individuals, firms and institutions to electronic analogue technology. He too points out that among many engineers there was an overt preference for electronic analogue computing. Edwards shows that electronic analogue computers offered several tangible advantages in comparison to the embryonic digital technology; i.e. greater reliability, smaller size, lower costs and the ease of data conversion when interfacing to analogue signals in the outside world. However, Edwards' focus is an examination of the roots of the digital computing "paradigm", he wants to reveal the extent to which it is a nexus of social, philosophical, political, military, economic and technical constructions. This he does very well indeed, and in a balanced way with respect to the analogue alternative. Most notably, he makes explicit the contemporary optimism towards the future of analogue computing: and to it co-existing with digital computing rather than being replaced by it.[12] Yet, his emphasis and argument is different to that of this book. Consequently, Edwards, does not examine in sufficiently great detail the analogue side of the story, including the scale and scope of the industry, the relationship between analogue computing and engineering culture and techniques, or the significance of the negotiations surrounding the definition of precision, accuracy and speed, or the importance of visualisation and model building and simulation in design practices. Furthermore, Edwards focuses almost entirely on the American context.

In the context of the existing literature on the history of computing, in this book I attempt to give a corrective or counter history. Indeed, what do we find if we travel down the "road not taken" (at least apparently not taken)?[13] We find a substantial international analogue computer community of developers and users. One that for many years held symposia and conferences, and developed and improved computing techniques as well as hardware performance. The community was evidently aware of, and in competition with, digital computing. But without the knowledge gleaned from hindsight, there was little reason for them to believe either that they would not be able to continue to improve their own technology, or that digital computer performance and reliability would necessarily continue to improve as prophesied. Viewing contemporary attitudes towards the future of computer development from the point of view of the alternative—the electronic analogue computer—reveals a much greater degree of scepticism and uncertainty about the ability of digital technology to deliver, as well as a much more robust belief in the future of analogue computing than portrayed by extant writings in the history of technology.

This book is an explicit critique of existing computer history. Here we are concerned with the incompatibility of the "tropes of progress" and success with that of contingency and analytical approaches which ignore or neglect the ever present alternatives, alternatives which arise from and are influenced by technical, cultural, political, economic choices and differences.

EXPLAINING TECHNOLOGICAL CHANGE: ALTERNATIVES TO THE UNILINEAR
SUCCESS STORY

For more than thirty years scholars have explicitly criticised work in the history of technology for its preoccupation with telling the stories of winners, either in the form of the heroic inventor or entrepreneur, or for accounts of artifacts and technological systems that read as "technological success stories". Much of this criticism has been directed at internalist accounts that take a dominant "successful" technology as their starting point and seek out precursors to it. As critics have pointed out, the resulting "straight-line narrative of increasing success" promotes a unilinear view of technical change that tends to obscure past alternatives and skew the historical account.[14]

The dominance of currently successful technologies and the bias towards studying successful technologies in historical and technical studies, has also tended to encourage a belief that the Darwinian theory of evolution is a valid and appropriate model for the processes of technological change. In this view, successful technologies are necessarily the fittest and best, having survived the rigours of a disinterested, multi-levelled process of evaluation and selection akin to natural selection.[15] In the relatively new field of evolutionary economics, it has been argued that particular patterns of development, or "trajectories", can be identified that are more "natural" than others.[16] Here again we see the influence of the analytical asymmetry towards success, since these "natural trajectories" invariably lead to dominant successful technologies; that is what makes them "natural". By emphasising the apparent necessity in the course of technical change, historical contingency and alternatives are simultaneously ruled out.

Yet studies show that the selection processes of technological change are not value-free, nor are technical merits necessarily determinant. We also know that alternatives to the existing dominant successful technologies either do exist or have existed and have subsequently disappeared. How then is the imbalance towards "success" and the general question of the existence and role of alternatives in technological change to be addressed, and how has it been addressed? Surveying the extant historical and more recent sociological literature on technology we see the co-existence of two distinct, but often overlapping, perspectives: the "failure studies" approach and the "symmetrical" approach.[17]

Failure studies

In practice the "failure studies" approach consists largely of the identification and study of a single technological failure, the dual goals being to explain why this particular technology failed and to counteract the dominance of linear success stories in the history of technology. One of the earliest and most explicit calls for a corrective study of "failure" came from Howard Mumford Jones. In 1959, in an article published in the first issue of the historical journal *Technology and Culture*, Jones—then Professor of English at Harvard University in the USA—argued:

> The history of technology is also a way of understanding man, but I suggest that the "failures" may be more illuminating than the "successes." For in what did failure lie? ... A history of failure is badly needed, but it must be a just and generous history, not a superior attitude towards human folly. Above all, the historian of technology must avoid the vulgar error of believing that nothing succeeds like success.[18]

In subsequent years in *Technology and Culture* similar criticisms and recommendations to those articulated by Jones were reiterated by Louis C. Hunter in the late 1960s, and by Robert Post, Eugene Ferguson and Reinhard Rurup in the early 1970s.[19] Yet we know from Staudenmaier's analysis of articles published in *Technology and Culture* between 1959 and 1980 that out of more than two hundred and seventy articles, only twenty one dealt with the subject of "technological failure".[20]

Staudenmaier explains the historian of technology's preoccupation with success in terms of the influence of a "progress myth" that has its origins in the "Scientific Revolution", and the growth of a belief in autonomous progress in scientific and technological domains. This progressive, linear, autonomous view of technical change has, according to Staudenmaier, been fostered largely by the internalist tradition in the history of technology. Internalist history has generally been written by participants. Typically, it takes the format of a fine grain analysis and descriptions of the development of either a currently "successful" and dominant technology, or of one of the precursors on a linear path leading to it. Though it is acknowledged that internalist accounts provide a valuable source of historically relevant data, they have been roundly criticised for ignoring contextual factors, and for implying that economic and political influences and human actions have neither a role, nor much explanatory power in technological change.[21]

Staudenmaier's analysis confirmed that until the early 1980s the hegemony of the success story had continued to go largely unchallenged. In the 1980s a small number of historians of technology continued to take a "failure studies" approach. One example is I. B. Holley Jr's 1987 study of the XP-75 Fighter Aircraft, the Ford Motor Company's war-time attempt at developing a mass-production aircraft. Holley introduces his article as follows: "Historians of technology have tended to write more often about successes than failures. This article investigates a significant failure—not to censure but to understand."[22] Larry Owens, in his 1986 study of the Rockefeller Differential Analyzer, built at MIT in the late 1930s and early 1940s, asks "What happened? Why did a twenty-one year effort to create a computer fail when it did?"[23]

More recently, a symposium on "Failed Innovations" was organised by the International Committee for the History of Technology (ICOHTEC).[24] In his introduction to the symposium, Hans-Joachim Braun reiterated the call to study failure and listed the benefits of the study of failure. He writes "In incorporating failed innovations into historical studies of technology, we can obtain a more realistic view of how technology developed." Braun continues:

> Dealing with failed innovations shows that an "internal" way of writing the history of technology is patently inadequate. As is known from cases already studied, the reasons for failure were, apart from technical deficiencies, often economic, social, political or cultural—or a mixture of some or all of them.[25]

Quite so, and clearly there is still a need and a role for a "history of failure". However, do the antonyms success and failure create an adequate analytical framework for the study of technical change? Moreover, how are "success" and "failure" defined? Though economists have developed criteria for defining success and failure in terms of markets, for historians difficulties arise in attempting to define technological success and failure for a number of reasons.[26] A technology that is successful in one context or application may fail in another and vice versa, thus raising the question, a failure or success for whom and for what? Technological success and failure have also been shown to be time-specific; successful technologies can later fail and failed technologies can later be successful. This is not to say that with the benefit of hindsight scholars cannot make a judgment as to which one of a number of competing technologies was more or less successful. But I suggest that if past alternatives are *a priori* described and analyzed either as "successes" or "failures" then much of the indeterminance and historical contingency of the processes of technical change may be lost.

Symmetrical studies

From the late 1970s the growing awareness and dissatisfaction with Whiggish success-story history, progress ideology, linear models of technological change and theories of autonomy of technology, prompted scholars to seek out new methodological approaches. Much of the subsequent methodological innovation was directed towards revealing the influence of "external" factors (including economic, social, political and cultural factors) on the direction of technical change. The principal goal was to unpack a rational, progressive, autonomous view of technology in order to empower individuals and society by emphasising that technology is socially shaped or socially constructed.[27] In connection with these goals, historians and sociologists developed approaches based on a symmetrical perspective that suspends *a priori* decisions as to the relative merits of alternative technologies. In this way, "success" and "failure" are studied and explained symmetrically by applying the same analytical methods to both.

David Noble's pioneering study *Forces of Production*, published in 1984, both deals with success and failure from the more general perspective of alternatives, and employs a symmetrical approach.[28] Noble's study examines the post-World War II development of automated machine tools in the USA. The alternative technologies in Noble's study are the record-playback method and numerical control. Noble argues that the decision by an alliance of military, industrial and academic institutions as to which alternative was to "succeed", was made not on the basis of technical superiority, but for political reasons grounded in self-interest. For Noble, employing a symmetrical approach involved "the reconstruction of lost alternatives" and served several purposes at the same time:

> First, it fills out the historical record and thus lays to rest the convenient fictions fostered by the ahistorical ideology of progress. Second, it reawakens us to a broader and largely available realm of possibilities. Third, it casts existing technologies in a new and critical light and thus stimulates reflection. Finally, and most important, such study of lost alter-

natives, which reveals the actual process of technological development, reveals also the patterns of power, cultural values, and the dominant ideas of the society which shaped that development.[29]

Very few symmetrical studies have actually been undertaken by historians, and none of these are as politically oriented as that of Noble. For the 19th century there is Robert Friedel's study of the development of Parkeside and Celluloid. Parkeside, introduced in the 1870s, was the first plastic, but its quality was unreliable and it was unsuccessful commercially. On the other hand, the manufacturers of the relatively more successful Celluloid paid much more attention to quality and careful marketing of their new product.[30] As far as the first half of the 20th century is concerned there is Eric Schatzberg's impressive study of the replacement of wood by metal in the construction of aeroplanes. Schatzberg argues that the technical evidence favoured neither alternative but that proponents of metal used rhetoric to link metal with progress and wood with stasis. Further that this "progress ideology of metal" ensured that a far greater proportion of resources went to improving metal aeroplanes.[31] Only Noble deals with a post-World War II technology.

The symmetrical explanation of success and failure and emphasis on controversy are key components of the analytical framework of the Social Construction of Technology (SCOT) programme. Social constructivism emerged in the mid-1980s.[32] Its origins lie in the Empirical Programme of Relativism in the sociology of scientific knowledge, which recommends taking scientific controversy as the principal focus of research.[33] In the sociology of science, symmetry implies that sociologists remain impartial in their analysis of conflicting claims of what is true or false. In the sociology of technology, a symmetrical analysis of controversy implies remaining impartial in debates on alternative technologies, and thus giving equal weight to claims of the superior performance of those technologies that fail and those that succeed. A feature common to the historical and SCOT approaches is the goal of analysing technology as a social phenomenon: both as social product and social force.

The social constructivist approach has been applied in a number of ways to a number of different technologies. Donald Mackenzie's book *Inventing Accuracy* is arguably the most skilfully written longer study employing this methodology to date. Mackenzie's book deals with the development of inertial guidance systems used in Intercontinental Ballistic Missiles, principally in the USA, but also in the former USSR.[34]

Comparing the subtitles of Noble's *Forces of Production: A Social History of Industrial Automation* and Mackenzie's *Inventing Accuracy: A Historical Sociology of Nuclear Missile Guidance* suggests that here we will find the ground on which the overlap between the study of technology as a historical and social phenomenon lies. Indeed, these studies do have much in common. Both studies are richly empirical, political, and take the American military industrial complex as their primary context. They both take a symmetrical approach to technological alternatives. Moreover, both are highly critical of uni-directional models of technical change, and of models that assert that technological change proceeds in a manner analogous to natural selection.

Noble argues that the social-Darwinian view of technical development leads to the assumption that all alternatives are always considered, and evaluated in a disinterested manner on the basis of technical merit and economic rationality. And thus that only the best survive. For Noble such "facile faith" in "objective science, economic rationality and the market", does not account for the actions of "people, power, institutions, competing values, or different dreams."[35]

Mackenzie's book attacks social-Darwinism of technology by providing a critical reassessment of the emerging "evolutionary" economics of technical change, championed by Nelson and Winter.[36] Briefly stated "evolutionary" models deal with technical change in terms of established technological "paradigms" and "regimes" that expand in various directions along trajectories. According to Nelson and Winter, those which appear "straighter" than others correspond to "natural trajectories".[37] In his book Mackenzie challenges the assertion that "natural trajectories" can be identified in the patterns of technological development and that it is a useful analytical metaphor. He points out "This book reveals just how wrong it is to assume that missile accuracy is a natural or inevitable consequence of technical change."[38]

Yet Mackenzie is not arguing there are no "trajectories", or that persistent patterns of technical change do not exist, nor—to use

Hughes's metaphor—that technological systems cannot exhibit "momentum" in a particular direction.[39] They can, but never momentum of their own: patterns persist because people invest time and money in trying to ensure that they do.[40] Indeed, both Noble and Mackenzie demonstrate that alternatives exist, each supported by its own institutions, or individuals who are committed to making their technology succeed. However, these individuals and institutions have different values, politics, preferences, and have uncertainties and expectations as to the future of their own and rival technologies. Thus, prior to one technology becoming dominant, separate or overlapping development paths may appear equally "natural" to the individuals involved. Furthermore, if the process by which one of these alternatives rose to dominance was, to quote from Noble, "objective science, economic rationality, and the market", there would be no, or only short technological controversies.[41] This raises the question "natural to whom and to what end?".

Returning to computer history, one could hardly be faulted for believing, both from the present-day digital computer perspective and the historical record, that the "natural trajectory" of postwar computer development was the digital computer "trajectory". But natural to whom? Having identified the successful digital computer as the "natural trajectory" one would probably, but not necessarily, have neglected the existence of an alternative electronic analogue computer tradition. However, subsequent analysis would most certainly not have revealed that the analogue computing community also saw their own technology as the "natural trajectory". And further that for them there was nothing "natural" about the eventual rise to dominance of digital technology. Any decision as to which path was truly *the* "natural trajectory" can only be made with hindsight and requires that alternatives and technological controversies be either ruled out or marginalised. Such ahistorical reconstruction is necessarily skewed by, and towards, the currently successful and dominant technology.

Studies, whether by historians, sociologists or economists, which acknowledge the existence and role of alternatives in technological change, point the way and offer the prospect of a more generous interpretation of history. More generous at least than Whig history and crude post-Kuhnian interpretations based on shifts from past paradigms to ever more successful ones.[42] The study of technological change from the point of view of alternatives (past and present) reveals the contingency of technical choices and can indeed "reawaken(s) us to a broader and largely available realm of possibilities".[43] Moreover, this study of the post-World War II analogue alternative demonstrates that a history generous enough to incorporate lost alternatives, is most certainly not a history of human follies but of human endeavours.

This study addresses issues in the history and sociology of technological change, and it has implications for a number of sub-disciplines in the history of technology. These issues and implications are discussed below.

Technological change and the military

Merritt Roe Smith has argued that though "military enterprise" has played a central role in the rise of the USA as a great industrial power, it remains the "least understood and appreciated phenomenon in American culture."[44] Smith defines "military enterprise" as consisting of a broad range of activities through which armed forces promote, coordinate and direct technological change. While adopting different analytical perspectives, Kenneth Flamm and Paul Edwards have already established the crucial role played by the military in founding and promoting the development of the electronic digital computer.[45] In this book, a major theme is the role of the military, both direct and indirect, in founding and sustaining a distinct electronic analogue computer industry. We see that the electronic analogue computer is an example of "military enterprise" working at many different levels. In the USA in particular, we see that the influence of military weapon programmes and funding for research and development was so pervasive, that in many respects electronic analogue computers were as much a military technology as the guided weapons, ICBMs and aircraft they helped to develop.

Consequently, this study also has implications for the history of post-World War II aviation and aerospace. Historians of aviation have largely ignored the role of computing in aeronautics. Yet electrical, electronic analogue and hybrid computers were used extensively in aeronautical engineering applications: for example, to model and simulate control systems, and for engine and airframe design. A history of aviation without a parallel history of the computational tools used, and an understanding of how these influenced design techniques and the direction of technical change, remains an incomplete one.[46]

Technology as knowledge; engineering and scientific cultures and communities

Electronic analogue computers were a product of engineering design, but also a tool that engineers used in the design of other technical systems. This study therefore addresses three interrelated issues in technical studies; technology as knowledge, the nature of engineering, and the scientisation of engineering. Eugene Ferguson, Edwin Layton and Walter G. Vincenti have in different ways made significant contributions to the debates on these subjects.

It is widely acknowledged that the view of the science–technology relation which portrays technology as applied science is no longer tenable. The applied-science model has been undermined by a growing consensus that views technology as highly differentiated forms of knowledge and cognition distinct from those in the scientific domain.[47] Layton, in his critical assessment of the technology as applied science model, argues "the separation of knowledge and technology is both recent and artificial. It is also self-contradictory. ... A common synonym for technology is 'know-how'. But how can there be 'know-how' without knowledge?" For a definition of technological knowledge Layton draws upon the Aristotelian view of technology as "systematic knowledge of the useful arts".[48] In Layton's view, technological knowledge can be divided into three categories: technological science (or theory), design, and technique. Yet here design is pivotal; according to Layton, design is "the critical act of synthesis in technology".[49]

The design perspective has been used not only to reveal the generation and use of technological knowledge but also to highlight differences in the practices and goals of engineering and science communities. Layton posits that they are "mirror-image twins" and that "'knowing' and 'doing' reflect the fundamentally different goals of the communities of science and technology."[50] Walter G. Vincenti argues "For engineers, in contrast to scientists, knowledge is not an end in itself or the central objective of their profession."[51] In the design-oriented view of Layton and Vincenti, while technological knowledge is generated during the design process it is also constrained to serve the needs of design. In science, knowledge is shaped by the needs of theory and thus to the construction of the most comprehensive and general theory possible.

To emphasise the uniqueness of technological knowledge, scholars have often stressed its tacit and visual nature. In his recent book *Engineering and the Mind's Eye*, Eugene Ferguson stresses the visual content of technological knowledge in relation to engineering practice and education.[52] He argues:

> Many features and qualities of the objects that a technologist thinks about cannot be reduced to unambiguous verbal descriptions; therefore, they are dealt with in the mind by a visual, nonverbal process. ... If we are to understand the nature of engineering, we must appreciate this important although unnoticed mode of thought.[53]

Moreover, commenting on the increasing scientisation of engineering education, Ferguson concludes:

> Necessary as the analytical tools of science and mathematics most certainly are, more important is the development in student and neophyte engineers of sound judgement and an intuitive sense of fitness and adequacy.[54]

In his analysis of problems arising out of the use of modern digital computers in engineering design, Ferguson draws on Henry Petroski's book *To Engineer is Human: The Role of Failure in Successful Design*.[55] The picture that emerges is of a community of modern engineers who have become preoccupied with numerical precision, and that they place too much faith in the correctness of computer-aided analysis programmes written by others and in the infallibility of results obtained from modern digital computers. Though apparently unaware of it, Ferguson echoes the arguments of a past generation of analogue computer "men", who stressed that a major advantage of electronic analogue computing over digital was that engineers were better able to develop "insight" and gain a "feel" for problem-solving. Moreover, that obtaining a good understanding of a problem was more important than the high numerical precision of computable data.

In his study of the use of parameter variation techniques in the design and testing of propellers, Vincenti also discusses the relevance of precision to engineering and science. He argues that a characteristic that can be used to distinguish between communities of engineers and scientists, is that the former are willing to forgo generality and precision to whatever degree the needs of the design dictate, while the latter value generality and precision for their own sake.[56]

On electronic analogue computers, data was generally represented graphically, they produced relatively low precision data, and were less flexible than electronic digital computers. As we shall see, they were pertinent to, and accepted and defended by an engineering culture that valued visualisation and an intuitive "feel" for problems over mathematical and analytical rigour, and "doing" over "knowing", even if this meant a loss of generality and precision. More generally, this study shows that in the post-World War II period, when engineering education placed more and more emphasis on mathematics and engineering science, differences and changes in the values and practices of engineering and scientific communities were made manifest through the design tools they developed, chose and used.[57]

STRUCTURE OF THE BOOK

There is a long tradition of developing and using analogue devices as aids to computation. Chapter 2 therefore begins with a brief historical survey of pre-20th century analogue computing devices. However, the bulk of the historical material in this chapter is concerned with developments in analogue computing between 1900 and 1945. One of the principal aims here is to indicate differences in the context and development of mechanical differential analyzers in the USA and Britain during the 1930s and early 1940s. In chapter 3, however, we see that the postwar electronic analogue computer was not a continuation of prewar work developing mechanical differential analyzers and AC network analyzers. By and large, the development of the electronic analogue computer was undertaken by different people working at different institutions on different problems. To familiarise the reader with postwar electronic analogue computing, chapter 3 also contains an outline of the major functional components of general-purpose electronic analogue computers and a description of how operational amplifiers can be configured to perform mathematical operations.

The principal aim of Chapters 4, 5 and 6 is to demonstrate the existence of a distinct postwar electronic analogue/hybrid computer industry in the USA and Britain, and to identify the major patterns of development and the influences of contextual factors and actors. Chapters 4 and 5 deal with the growth and decline of electronic analogue computing in the USA. Chapter 3 covers the period from the late 1930s to the mid-1950s, and with some overlap chapter 4 continues the story from the mid-1950s to the late-1970s.

Chapter 4 begins with an account of the origins of electronic analogue computing. In this chapter we also see the formative role played by funding from the Office of Naval Research and the US Air Force in transforming bench-top, small-scale experimental systems into full-scale commercial electronic analogue computers. Military influence on the growth of a distinct postwar analogue computer industry operated at several levels, including the direct funding of analogue computer R&D, diffusion of the technology and techniques, as direct customer, and sponsor of R&D in private industry, for which firms subsequently purchased analogue computers.

In chapter 5 (on the USA) and chapter 6 (on Britain) the bulk of the analysis is at the level of the firm. Three analytical categories are defined: the developer/users, user/manufacturers and non-user/manufacturers. We see that the general pattern is from firms that developed analogue computers for their own use, through an intermediate stage, towards the dominance of specialised firms that developed and manufactured, but did not use analogue and hybrid computers themselves.

We see that by the mid-1950s more than a dozen firms were manufacturing general-purpose electronic analogue computers on a commercial basis. Details of these firms and their products are given. In addition, many firms, military and government R&D and test facilities designed and constructed systems for use in their own laboratories. Though data has been gathered on many of these, the degree of similarity exhibited means that cataloguing each and every one adds little to the account. Therefore, a few of historical interest have been selected to illustrate the range and degree of involvement of research organisations, private firms and military laboratories in the development and use of electronic analogue computing equipment.

In chapter 5 we see that, in the USA, by the early 1950s the electronic analogue computer had become established as a commercially viable and manufacturable product. The responsibility and financial burden for subsequent R&D gradually shifted from military-funded projects to the analogue computer firms. We see that in the USA from the mid-1950s the formative period characterised by many developer/user firms draws to a close. The availability of commercial analogue computers means that it made little sense for firms to commit resources to re-inventing electronic analogue computer. However, we see that computational demands arising from the development of ICBMs resulted in the user-led development of hybrid computers. The first experimental hybrid computers were developed by firms in the aeronautics industry and combined separate analogue and digital computers, with the aim of surmounting the computational shortcomings of each. At the time this was not a direction being pursued by the analogue computer firms. The leading firms were beginning to incorporate digital methods for the purpose of setting-up and control of the analogue, but not for shared computation. However, these firms faced stiff competition from faster, increasingly user-friendly digital computers, and the complexity and size of computational problems continued to grow. The analogue computer firms responded by developing a wide variety of hybrid computer systems that married the computational, data-handling characteristics and programmability of the digital computer with the computational speed of the analogue computer. Towards the end of chapter 5 we see that even though hybrid computers offered considerable economic benefits in particular applications, two factors worked in the favour of all-digital computing: the additional skills required to operate a hybrid computer, and the wide-spread adoption of digital computers.

This book is a comparative one. In chapter 6 I describe and analyze the growth of electronic analogue computing in the United Kingdom from 1940 to the mid-1970s, and draw comparison with developments in the USA. We see that in Britain the principal context for the initial development of electronic analogue computers was in association with guided missiles and military and civil aircraft design. Much of the development work grew out of wartime research and took place in research laboratories owned and funded by the British Government, the foremost being the Royal Aircraft Establishment at Farnborough. We see that in Britain aeronautics was not the only significant or influential context for electronic analogue computers. Research and development associated with electric power generation and nuclear reactor design led to the development, construction and purchase of analogue computer installations, large and small. The relative importance of applications outside of the aeronautics industries in Britain was in contrast to the situation in the USA. This can be explained largely in terms of the enormous scale and scope of the American ICBMs and manned space flight programmes, and the near-absence of similar programmes in Britain. In Britain the postwar commercialisation of electronic analogue computers began in 1953, and independent British firms continued to develop and manufacture analogue and hybrid computers until 1970. As we will see, by 1953 commercial systems had already been available in the USA for several years, the early lead taken by the Americans, especially in hardware development, was maintained throughout the 1950s, '60s and '70s. By the early 1960s American firms had established a strong presence in the British market and by 1970 had almost complete control.

In chapter 7 we see a change in emphasis, and I explore analogue computing in the context of engineering culture and design. I argue that electronic analogue computers were not only shaped by, but also embodied and extended a 19th and early 20th century engineering pedagogy, characterised by the use of graphical methods, scale model building, as well as empirical trial-and-error and parameter variation methods. We see that the increasing size and complexity of technical systems, and thus problems in engineering design, worked not only to stimulate the development and use of electronic analogue computers, but latterly also to undermine their usefulness.

Chapter 8 analyzes the controversy between advocates of analogue and digital computing. We see that though the terms of the debate appeared to be overtly technical and quantitative, they were in fact subject to negotiation, redefinition and reinterpretation. There was also an explicitly qualitative aspect of the debate which stressed the interactive nature of analogue computing, its one-to-one correspondence with the real-world systems under study, and the feel and insight which analogue computing gave users into the problems that they were investigating. We see that for the analogue computer community, the future dominance of digital computing was not regarded as a foregone conclusion. The analogue community recognised that the digital computer was a competitive technology and that it had particular advantages in particular applications. However, it did not represent the "one best way forward" for computer development. Nevertheless, the analogue community did have to counter the progressive rhetoric and wonder myths that built up around the digital computer. Consequently, the analogue community invoked its own rhetoric and sought to maximise what I refer to as "application domains" for analogue/hybrid computers. Chapter 9 is the concluding chapter.

SOURCES

Historical data for this research have come from a wide range of sources, including text books, professional and trade journals, conference reports, newspapers, magazines, publications by independent research organisations, internal and external publications by military and government research organisations, trade literature and internal documents generated by the analogue computer companies, personal correspondence and the minutes of a user group. Much of this material came from archive-based research undertaken at the following locations: in Britain at the National Archive for the History of Computing, Manchester, the Institute of Electrical Engineers, London, the Science Museum, London; and in the USA at the Charles Babbage Institute Minneapolis, IEEE Centre for the History of Electrical Engineering, New York, the Smithsonian Air and Space Museum, Washington DC, the National Museum of American History, Washington DC, the United States National Archive, Washington DC, the Computer Museum, Boston, Ma., the AT&T Archives, Warren NJ., the Hagley Museum and Library, Wilmington, Del., and MIT Archive Centre, Cambridge, Ma.

In addition to use of textual sources I have also interviewed and had discussions with several pioneers and others who worked in the analogue/hybrid computer industry and/or used them. The experiences of these individuals cover a variety of different aspects of analogue/hybrid computing: from design, development, manufacture and sale to their use in industrial and academic applications.

NOTES

1.   A. Hass, "Gigantic Computer Goes to Junk Pile", *Philadelphia Inquirer*, Philadelphia, Pa., 19 Sept 1968.

2.   M. C. Mahoney, "The history of computing in the history of technology", *Annals of the History of Computing*, Vol. 10, no. 2, 1988, pp. 113–125, p. 114. See also W. Aspray, "The History of Computing Within the History of Information Technology", *History and Technology*, Vol. 11, no. 1, 1994, pp. 7–19. And for a comprehensive guide to the literature see, J. Cortada, *A Bibliographical Guide to the History of Computing, Computers, and the Information Industry*, Greenwood Press, Westport Conn., 1990.

3.   This remains true even for relatively recent studies. See A. Goldberg, (ed.), *A History of Personal Workstations*, Addison-Wesley, Reading, Mass., 1988; D. E. Lundstrom, *A Few Good Men From Univac*, MIT Press, Cambridge, Mass., 1987; S. Nash, (ed.), *A History of Scientific Computing*, Addison-Wesley, Reading, Mass., 1990; E. W. Pugh, *Memories That Shaped an Industry*, MIT Press, Cambridge, Mass., 1984.

4.   See W. Aspray, (ed.), *Computing Before Computers*, Iowa State University Press, Ames, Iowa, 1990; W. Aspray, *John von Neumann and the Origins of Modern Computing*, MIT Press, Cambridge, Mass., 1990; P. E. Ceruzzi, *Reckoners: The Prehistory of the Digital Computer, From Relays to the Stored Programme Concept, 1935–1945*, Greenwood Press, Westport, Conn., 1982; M. Campbell-Kelly, *ICL: A Business and Technical History*, Clarendon, Oxford, 1989; N. Stern, *From ENIAC to UNIVAC, An Appraisal of the Eckert-Mauchly Computers*, Digital Press, Bedford, Mass., 1981; M. Williams, *A History of Computing Technology*, Prentice-Hall, Englewoods Cliffs, NJ, 1985.

5.   For economic and business history see N. Dorfman, *Innovation and Market Structure: Lessons from the Computer and Semiconductor Industries*, Ballinger Books, Cambridge, Mass., 1987; K. Flamm, *Targeting the Computer: Government Support and International Competition*, The Brookings Institution, Washington, DC, 1987; K. Flamm, *Creating the Computer: Government, Industry, and High Technology*, The Brookings Institution, Washington, DC, 1988; J. Hendry, *Innovating for Failure*, MIT Press, Cambridge, Mass., 1989. For studies taking a socio-historical perspective see J. van den Ende, *The Turn of the Tide: Computerization in Dutch Society 1900–1965*, Delft University Press, Delft, 1994; P. N. Edwards, *The Closed World: Computers and the Politics of Discourse in Cold War America*, MIT Press, Massachusetts, 1996; B. Elzen and D. Mackenzie, "From Megaflops to Total Solutions", in W. Aspray, (ed.), *Technological Competitiveness: Contemporary and Historical Perspectives on the Electrical, Electronics, and Computer Industries*, IEEE Press, Piscataway, NJ, 1993; J. S. Small, "General-purpose Electronic Analog Computing: 1945–1965", *Annals of the History of Computing*, Vol. 15, 1993, pp. 8–18; J. S. Small, "Engineering, Technology and Design: The Post-Second World War Development of Electronic Analogue Computers", *History and Technology*, Vol. 11, no. 1, 1994, pp. 33–48; J. S. Small, "Electronic Analog Computer", in *Instruments of Science: an Historical Encyclopedia*, Garland Publishing, New York, 1998. See the special issue of the *Annals of the History of Computing*, on analogue computing techniques—mainly differential analyzers: J. M. Nyce, (ed.), *Annals of the History of Computing*, Vol. 18, no. 4, 1996. See also the journalistic account by J. T. Kidder, *The Soul of a New Machine*, Atlantic-Brown, Boston, Mass., 1981; Special Issue on Analogue Computers.

6.   W. Aspray, "Edwin L. Harder and the Anacom: Analogue Computing at Westinghouse", *Annals of the History of Computing*, Vol. 15 no. 2, 1993, pp. 35–52; P, Ceruzzi, *Beyond the Limits: Flight Enters the Computer Age*, MIT Press, Mass., 1989; P. Ceruzzi, "Electronics Technology and Computer Science, 1940–1975: A Coevolution", *Annals of the History of Computing*, Vol. 10 no. 4, 1989, pp. 257–275; A. B. Clymer, "The Mechanical Analogue Computers of Hannibal Ford and William Newell", *Annals of the*

History of Computing, Vol. 15 no. 2, 1993, pp. 19–34; L. Hassig (ed.), *Alternative Computers*, part of the *Understanding Computers Series*, Time-Life Books, Alexandria, Virginia, 1989; P. A. Holst, "George A. Philbrick and Polyphemus", *Annals of the History of Computing*, Vol. 4 no. 3, 1982, pp. 143–156; J. McLeod, (ed.), *Pioneers and Peers*, The Society for Computer Simulation International, San Diego, 1988, pp. 4–10; J. E. Tomayko, "Helmut Hoelzer's Fully Electronic Analogue Computer", *Annals of the History of Computing*, Vol. 7 no. 3, 1985, pp. 227–240.

7. See, S. Augarten, *Bit by Bit: An Illustrated History of Computers*, George Allen & Unwin, London, 1985; C. Evans, *The Making of the Micro: A History of the Computer*, Victor Gollancz, London, 1981; H. H. Goldstine, *The Computer: from Pascal to von Neumann*, Princeton University Press, Princeton, 1972; S. H. Lavington, *Early British Computers: The Story of Vintage Computers and the People Who Built Them*, Manchester University Press, Manchester, 1980; M. Williams, *A History of Computing Technology*, Prentice-Hall, Englewood Cliffs, NJ, 1985.

8. M. Campbell-Kelly and W. Aspray, *Computer: A History of the Information Machine*, Basic Books, New York, 1996.

9. Ibid., p. 2

10. P. N. Edwards, *The Closed World: Computers and the Politics of Discourse in Cold War America*, MIT Press, Massachusetts, 1996; Also for an alternative approach to the history of analogue and digital computing see, A. Tympas, "From Digital to Analogue and Back: The Ideology of Intelligent Machines in the History of the Electrical Analyzer, 1870s-1960s, *Annals of the History of Computing*, Vol. 18, no. 4, 1996, pp. 42–48; also for a short non-contextualized account see the chapter by A. G. Bromley, "Analogue Computing Devices", pp. 156–199, in W. Aspray, (ed.), *Computing Before Computers*.

11. P. N. Edwards, *The Closed World*, p. xii.

12. P. N. Edwards, *The Closed World*, see chapter 2, pp. 66–73.

13. D. F. Noble, *Forces of Production: A Social History of Industrial Automation*, Victor Gollancz, New York, NY, 1984.

14. H. M. Jones, "Ideas, history, technology", *Technology and Culture*, Vol. 1, no. 1, 1959, pp. 20–27, p. 24. See also, J. M. Staudenmaier, *Technology's Storytellers: Reweaving the Human Fabric*, MIT Press, Cambridge, Mass., 1985, pp. 175–176; E. S. Ferguson, "Towards a discipline of the history of technology" *Technology and Culture*, Vol. 15, no. 1, 1974, pp. 13–24.

15. D. F. Noble, *Forces of Production*.

16. R. R. Nelson and S. G. Winter, *An Evolutionary Theory of Economic Change*, Belknap/Harvard University Press, Cambridge, Mass., 1982. see chapter 11.

17. The term "failure studies" was coined by Staudenmaier as a classification for a range of studies that either discussed the success/failure imbalance and/or gave accounts of failure in technological change. J. M. Staudenmaier, "What SHOT Hath Wrought and What SHOT Hath Not: Reflections on Twenty-five Years of the History of Technology", *Technology and Culture*, Vol. 25, no. 3, 1984, pp. 707–730.

18. H. M. Jones, "Ideas, history, technology", *Technology and Culture*, p. 25.

19. E. S. Ferguson, "Towards a discipline of the history of technology"; L. C. Hunter, "The living past in the Appalachias of Europe: Water-mills in southern Europe", *Technology and Culture*, Vol. 8, no. 4, 1967, pp. 446–466; R. C. Post, "The Page Locomotive: Federal sponsorship of invention in mid-19th-century America", *Technology and Culture*, Vol. 13, no. 2, 1972, pp. 140–169; R. Rurup, "Historians and modern technology: Reflections on the development and current problems of the history of technology", *Technology and Culture*, Vol. 15, no. 2, 1974, pp. 161–193.

20. J. M. Staudenmaier, *Technology's Storytellers: Reweaving the Human Fabric*, pp. 175–176.

21. Ibid.

22. I. B. Holley Jr., "A Detroit dream of mass-produced fighter aircraft: The XP-75 fiasco", *Technology and Culture*, Vol. 28, no. 3, 1987, pp. 578–593, p. 578.

23. L. Owens, "Vannevar Bush and the Differential Analyzer: The Text and Context of an Early Computer", *Technology and Culture*, Vol. 27, no. 1, 1986, pp. 63–95, p. 65.

24. The symposium on "Failed Innovations" was organized by ICOHTEC (International Committee for the History of Technology) and was held at the International Congress of History of Science in Hamburg and Munich Aug 1–9, 1989. Eleven papers were subsequently published in *Social Studies of Science* Vol. 22, no. 2, May 1992. R. A. Buchanan, "The Atmospheric Railway of I. K. Brunel", *Social Studies of Science*, Vol. 22, no. 2, May 1992, 231–244; H. S. Torrens, "A Study of 'Failure' with a 'Successful Innovation': Joseph Day and the Two-Stroke Internal-Combustion Engine", *Social Studies of Science*, Vol. 22, no. 2, May 1992, 245–262; E. N. Todd, "Electric Ploughs in Wilhelmine Germany: Failure of an Agricultural System," *Social Studies of Science*, Vol. 22, no. 2, May 1992, 263–282; M. Efmertová, "Czech Physicist Jaroslav Safránek and his Television", *Social Studies of Science*, Vol. 22, no. 2, May 1992, 283–300; W. D. Lewis and W. F. Trimble, "The Airmail Pickup System of All American Aviation: A Failed Innovation?", *Social Studies of Science*, Vol. 22, no. 2, May 1992, 301–316; A. N. Stranges, "Farrington Daniels and the Wisconsin Process for Nitrogen Fixation", *Social Studies of Science*, Vol. 22, no. 2, May 1992, 317–338; Hans-Joachim Braun, "The Chrysler Automotive Gas Turbine Engine, 1950–80, *Social Studies of Science*, Vol. 22, no. 2, May 1992, 339–352; R. T. McCutcheon, "Science, Technology and the State in the Provision of Low-Income Accommodation: The Case of Industrialized House Building, 1955–77", *Social Studies of Science*, Vol. 22, no. 2, May 1992, 353–372; J. Hult, "The Itera Plastic Bicycle", *Social Studies of Science*, Vol. 22, no. 2, May 1992, 373–386; B. C. Hacker, "The Gemini Paraglider: A Failure of Scheduled Innovation, 1961–64", *Social Studies of Science*, Vol. 22, no. 2, May 1992, 387–406.

25. Hans-Joachim Braun, "Introduction [to Symposium on 'Failed Innovations']", *Social Studies of Science*, Vol. 22, no. 2, May 1992, 213–230, p. 214.

26. Economists have developed a taxonomy of success and failure based on stages from invention, manufacture, marketing and diffusion of commercial products. See C. Freeman, *The Economics of Industrial Innovation*, Frances Pinter, London, 1982.

27. See, W. E. Bijker, T. P. Hughes and T. Pinch, (eds.), "*The Social Construction of Technological Systems: New Directions in the Sociology and History of Technology*", MIT Press, Cambridge, Mass., 1987; R. S. Cowan, *More Work for Mother*, Basic Books, New York, NY, 1983; T. P. Hughes, *Networks of Power, Electrification in Western Society, 1880–1930*, Johns Hopkins University Press, Baltimore, 1983; D. Mackenzie and J. Wajcman, (eds.), *The Social Shaping of Technology*, Open University Press, Milton Keynes, 1985; L. Winner, *Autonomous Technology*, MIT Press, Cambridge, Mass., 1977.

28. D. F. Noble, *Forces of Production*.

29. Ibid., p. 146.

30. R. Friedel, "Parkeside and Celluloid: The Failure and Success of the First Modern Plastic", *History and Technology*, Vol. 4, no. 1, 1979, pp. 45–62.

31. See E. Schatzberg, *Ideology and Technical Change: The Choice of Materials in American Aircraft Design Between the World Wars*, PhD thesis, University of Pennsylvania, 1990, published as E. Schatzberg, *Wings of Wood, Wings of Metal: Culture and Technical Choice in American Airplane Materials, 1914–1945*, Princeton University Press, Princeton, 1999; E. Schatzberg, "Ideology and Technical Choice: The Decline of the Wooden Airplane in the United States, 1920–1945", *Technology and Culture*, Vol. 35, no. 1, 1994, pp. 34–69.

32. T. J. Pinch and W. E. Bijker, "The Social Construction of Facts and Artifacts: or How the Sociology of Science and the Sociology of Technology Might Benefit Each Other", *Social Studies of Science*, Vol. 14, 1984, pp. 399–441; W. E. Bijker, T. P. Hughes and T. Pinch, (eds.), *The Social Construction of Technological Systems*.

33. H. Collins, "Stages in the Empirical Programme of Relativism", *Social Studies of Science*, Vol. 11, 1981, pp. 3–10.

34. D. Mackenzie, *Inventing Accuracy: A Historical Sociology of Nuclear Missile Guidance*, MIT Press, Cambridge, Mass., 1990. For a critical assessments of this book and the programme of the social constructivist's programme, see the review essay by D. Edgerton, "Tilting at Paper Tigers", *British Journal for the History of Science*, Vol. 26, no. 88, March 1993, pp. 67–75; and L. Winner, "Social Constructivism: Opening the Black Box and Finding it Empty", *Science as Culture*, Vol. 3, no. 16, pt. 3, 1993.

35. D. F. Noble, *Forces of Production*, pp. 144–145.

36. R. R. Nelson and S. G. Winter, *An Evolutionary Theory of Economic Change*, Belknap/Harvard University Press, Cambridge, Mass., 1982. see chapter 11.

37. Ibid., p. 258; see also G. Dosi, "Technological Paradigms and Technological Trajectories: A Suggested Interpretation of the Determinants of Technical Change", *Research Policy*, Vol. 11, 1982, pp. 147–162.

38. D. MacKenzie, *Inventing Accuracy*, p. 3.

39. T. P. Hughes, "Technological Momentum in History: Hydrogenation in Germany, 1898–1933", *Past and Present*, No. 44, Aug 1969, pp. 106–132; T. P. Hughes, "The Evolution of Large Technological Systems", in W. E. Bijker, T. P. Hughes and T. Pinch, (eds.), "*The Social Construction of Technological Systems*, pp. 51–82.

40. D. MacKenzie, *Inventing Accuracy*, p. 165–169.

41. D. F. Noble, *Forces of Production*, p. 144.

42. T. S. Kuhn, *The Structure of Scientific Revolutions*, 2nd ed., Chicago University Press, Chicago, 1970; see also E. W. Constant II, *The Origins of the Turbojet Revolution*, Johns Hopkins University Press, Baltimore, 1980.

43. D. F. Noble, *Forces of Production*, p. 146.

44. M. R. Smith, (ed.), *Military Enterprise and Technological Change: Perspectives on the American Experience*, MIT Press, Cambridge, Mass., 1985, p. 4.

45. K. Flamm, *Creating the Computer: Government, Industry, and High Technology*, The Brookings Institution, Washington, DC, 1988; P. N. Edwards, *The Closed World*, 1996

46. The co-development of computers and post-World War Two aviation and aerospace has however been addressed by a computer historian, see P. Ceruzzi, *Beyond the Limits: Flight Enters the Computer Age*, MIT Press, Cambridge, Mass., 1989.

47. See, B. Barnes, "The Science-Technology Relationship: A Model and A Query", *Social Studies of Science*, Vol. 12, 1982, 166–172; R. Laudan, (ed.), *The Nature of Technological Knowledge: Are Models of Scientific Change Relevant?*, D. Reidel, Dordrecht, 1984.

48. E. T. Layton Jr., "Technology as Knowledge", *Technology and Culture*, Vol. 15, no. 1, 1974, pp. 31–41, pp. 33–34.

49. E. T. Layton, Jr., "Through the Looking Glass, or News from Lake Mirror Image", *Technology and Culture*, Vol. 29, no. 3, 1987, pp. 594–607, p. 604.

50. E. T. Layton, Jr., "Technology as Knowledge", p. 40; E. T. Layton, Jr, "Mirror Image Twins: The Communities of Science and Technology in Nineteenth Century America", *Technology and Culture*, Vol. 12, no. 4, 1974, pp. 562–580; this paper also appears in, G. H. Daniels (ed.), *Nineteenth Century American Science: A Reappraisal*, Northwestern University Press, Evanston, Ill., 1972.

51. W. G. Vincenti, *What Engineers Know and How They Know It: Analytical Studies from Aeronautical History*, Johns Hopkins University Press, Baltimore, 1990, p. 6.

52. E. S. Ferguson, *Engineering and the Mind's Eye*, MIT Press, Cambridge, Mass., 1992; see also E. S. Ferguson, "The Mind's Eye: Nonverbal Thought in Technology", *Science*, Vol. 197, 1977, pp. 827–836.

53. E. S. Ferguson, *Engineering and the Mind's Eye*, "Preface".

54. Ibid., p. 194.

55. H. Petroski, *To Engineer is Human: The Role of Failure in Successful Design*, Macmillan, London, 1985.

56. W. G. Vincenti, "The Air-Propeller Test of W. F. Durand and E. P. Lesley: A Case Study in Technological Methodology," *Technology and Culture*, Vol. 20, no. 4, 1979, pp. 712–751; see also W. G. Vincenti, "Control-Volume Analysis: A Difference in Thinking Between Engineering and Physics," *Technology and Culture*, Vol. 23, no. 1, 1982, pp. 145–174; W. G. Vincenti, "The Davis Wing and the Problems of Airfoil Design: Uncertainty and Growth in Engineering Knowledge," *Technology and Culture*, Vol. 27, no. 4, 1986, pp. 717–758.

57. B. Seely, "Research, Engineering, and Science in American Engineering Colleges: 1900–1960", *Technology and Culture*, Vol. 34, no. 2, pp. 344–386.

CHAPTER 2

Analogue Computing Devices in the 19th and Early 20th Centuries

INTRODUCTION

This chapter has two main sections. The first is an introduction to analogue computing devices, their classification, and how they differ from digital/numerical machines. The second begins with a brief historical survey of analogue devices developed in the 19th century. This is followed by a review of analogue devices that emerged in the first half of the 20th century. However, this section is primarily concerned with the context and construction of the mechanical differential analyzers in the USA and Britain in the 1930s and early 1940s.

Analogue devices have been in use since ancient times. They have been used to perform many different functions and given many forms: mechanical, electrical, hydraulic, and so forth. Several of the analogue devices invented prior to the 20th century are relatively well known, such as the sundial, astrolabe and slide rule, while others such as the harmonic analyzers, tide predictors and integraphs are now quite obscure. In the 20th century, the tradition of the use of analogy as the fundamental principle of operation in scientific instruments and calculation devices continued. Indeed, an enormous variety of analogue devices were developed in that century. Many of them were highly specialised devices developed for the solution of one particular problem. Others were designed to be more versatile, and capable of being used in a wider variety of applications. Among the most noteworthy devices in the latter category are network analyzers, mechanical differential analyzers and electronic analogue computers.

ANALOGUE COMPUTING DEVICES: CLASSIFICATION AND CHARACTERISATION

The unifying characteristic of analogue computing devices is that they are continuous mathematical machines. Data in analogue devices varies continuously, is operated upon continuously and transferred around the machine and from input to output in a continuous form. This is the principal feature that differentiates them from digital computing devices which are fundamentally arithmetic machines and operate in discrete steps on discontinuous data.[1]

These differences can perhaps be better understood by comparing the operation of the abacus with that of the slide rule. In the abacus, quantities are represented by a number of beads, thus the quantity being represented can only vary, up or down, by a minimum of one bead—there are no partial beads. Furthermore, the abacus is an arithmetic machine, in which all operations are performed as a series of additions or subtractions. In contrast, slide rules represent quantities as continuously varying magnitudes: in this case length. Results are obtained by adding or subtracting measured lengths. The granularity of the result is limited only by the coarseness of the scale used to perform the measurement. Though in practice the slide rule worked by adding and subtracting lengths, they were fundamentally mathematical devices which operationalised logarithms: a mathematical function that was used to facilitate multiplication and division.

Another example of the use of length as the analogous magnitude is in the use of scale drawings and scale models. Yet though the slide rule and physical scale models/drawings fall under the rubric of analogue computing devices, they belong to two distinct yet overlapping categories: the indirect (or functional) and direct analogue devices. Table 2.1 below illustrates, by the use of a few examples, how the wide spectrum of analogue devices can be subdivided into these two categories.

Direct analogues have been defined historically as those devices which, at the level of each component of the analogue model, have a one-to-one correspondence with the physical systems under investigation. Thus parameters and variables in one system correspond to those of the analogous system. Scale drawings, physical scale models and network analyzers are among the foremost examples of direct analogues. Much of the appeal of problem-solving by direct analogy was that problems could be solved without recourse to abstract

TABLE 2.1 DIRECT AND INDIRECT ANALOGUE COMPUTING DEVICES

Analogue Devices and Computers				
Indirect/Functional			**Direct**	
Electrical/Electronic	Mechanical	Electrical	Mechanical	Hydraulic Pneumatic Acoustic
Flight simulator	Slide rule	AC Network analyzer	Scale models	Scale models
Process controller	Gun director	DC Network analyzer	Wind tunnels	
Electronic analogue computer	Mechanical differential analyzer	Electrolytic tank		
		Conducting paper		

mathematical techniques. Moreover, structures or dynamic systems could be modeled and tested without ever having to fully derive the equations that described them.

The category of indirect analogue devices has also been referred to as "functional" analogue devices.[2] This latter description is more illuminating. It refers to the fact that devices in this category are constructed from analogue computing elements that are designed to perform specific mathematical *functions*, such as integration or trigonometrical functions. These analogue computing elements ranged from relatively simple mechanical linkages to complex cams, and from wheel-and-disk integrators to electronic operational amplifiers.

The starting point for the construction of a functional analogue device was generally to derive the equations that defined the system being studied. Analogue devices were then designed to operationalise the equations and automate their solution. The functional analogue thus consisted of a combination of mechanical, electrical or electronic computing elements, each constrained to perform a specific mathematical or arithmetic operation.

Up to the 20th century the majority of analogue computing devices, both direct and functional, were designed to perform either one specific calculation or a specific class of mathematical problems.

These devices have generally been referred to as single, fixed, or more commonly special-purpose devices. The inherently close correspondence that is fundamental to direct analogy has meant that direct analogue computing devices were almost exclusively special-purpose devices, the chief exception being the network analyzers, the first of which was developed in the 1920s. Though developed primarily for the analysis of electrical transmission and distribution systems, network analyzers were also used to model mechanical systems and to solve heat transfer and fluid flow problems.[3] Planimeters, integraphs and harmonic analyzers are among the foremost examples of the functional analogue devices developed in the 19th century that were special purpose in terms of their application. Gun directors and range finders, such as those designed by Hannibal Ford and William Newell for the US Navy during the first half of the 20th century, are further examples of special-purpose functional analogue devices.[4]

Yet not all analogue computing devices were regarded as special purpose. There are a few that by virtue of their versatility have been classified as "general purpose"—though not without some controversy.[5] Furthermore, general-purpose devices have almost without exception been functional analogue computing devices. For example, the slide rule, a functional analogue computing device, was a versatile aid to computation. Invented in the seventeenth century by William Oughtred, the slide rule was the "pocket calculator" of engineering and science until the late 1970s.[6] Yet though there are other examples, the classification "general purpose" has usually been reserved for two 20th century inventions: the mechanical differential analyzer and the electronic analogue computer.

The first large-scale general-purpose analogue computer was the mechanical differential analyzer developed in the late 1920s by Vannevar Bush at the Massachusetts Institute of Technology in the USA. The differential analyzer differed from previous devices in its scale and scope. Earlier devices could solve first-order differential equations directly, or second-order differential equations by indirect methods. Bush's differential analyzer was capable of solving a sixth-order differential equation or two third-order differential equations simultaneously. The differential analyzer was a generalised equation-solver. By altering the arrangement and interconnection of mechanical computing elements, the differential analyzer could be set up (or programmed) to solve a wide variety of problems in engineering and science.

The general-purpose electronic analogue computer was a post-World War II invention. Commercial systems were normally modular in design and could be combined for more complex problems. Even a small desk-top system could solve sixth-order differential equations. A large installation, consisting of approximately two hundred and fifty computing amplifiers and auxiliary equipment could, among other things, solve a fiftieth-order differential equation. Fundamentally, the electronic analogue computer was a functional analogue device. However, they were also often employed in the manner of a direct analogue computing device and thus exhibited a degree of duality between functional and direct computing techniques which was central to their versatility. This duality corresponds to two of the principal ways in which electronic analogue computers were used: as differential analyzers and as model-building and simulation systems.

When using the electronic analogue computer as a differential analyzer the starting point was a set of equations. To solve these equations, a corresponding functional analogue was built by combining computing elements. When being used to construct electronic models of dynamic physical systems for the purpose of simulation, the model could be constructed without defining the system mathematically. In such cases the electronic analogue computer was being employed as a direct analogue device.

ANALOGUE COMPUTING DEVICES FROM THE LATE 19TH CENTURY TO THE MID-20TH CENTURY: A SURVEY

Prior to the 19th century most analogue computing devices were designed to solve mathematical problems defined by geometric relationships. However, the majority of the mechanical analogue computing devices invented in the 19th century were designed to operationalise the mathematical functions of integral calculus. Many of these were based on a device designed, though never constructed, by a Bavarian engineer, J. H. Hermann, in 1814. Known as a "planimeter", this device used the principles of integral calculus to measure the area of small irregularly shaped surfaces, such as the scale drawing of a piece of land. Hermann's planimeter design incorporated a small wheel-and-disk mechanical integrator, consisting of two disks between which the wheel was held under compression. In the 19th century much of the work to improve integrator design came about as a result of attempts to improve the overall performance of planimeters. One of the most successful designs to emerge was the polar planimeter inverted in 1854 by J. Amsler. More than 12,000 Amsler planimeters were sold in the 30 years following its introduction.[7]

The 19th century also saw the development of the first direct electrical analogue techniques: the electrolytic tank and conducting paper. These techniques were designed as aids to computation in the solution of boundary value problems, such as temperature and magnetic field distribution. The electrolytic tank and conducting paper operated on a principle known as *field analogy*. The field analogy method is based on the analogy between the pattern of distribution of an electric field and, for example, the pattern of distribution of temperature across a conducting medium. The first publication of a field plot derived experimentally by this technique appeared in a paper by G. Kirchhoff in 1845. The conducting paper method was performed by measuring and marking the points of equipotential between two electrodes on the surface of the paper; drawing in the line to connect these points gave the two-dimensional field distribution pattern.[8]

The electrolytic tank was a more widely used technique. A model of the system under investigation was placed in a tank of mild electrolyte. The model formed one electrode, and a thin wire probe the other. The space around the model was then systematically explored with the probe to determine the field distribution pattern in two dimensions.[9] In Britain in 1876 W. G. Adams published the first description of the use of this technique in the solution of field equations in the *Proceedings of the Royal Society*.[10]

In the same year that Adams published his account of the electrolytic tank technique, two papers of great importance to the history of mechanical analogue computing devices were also published in the *Proceedings of the Royal Society*. The authors were Professor James Thomson and his brother William Thomson—later Lord Kelvin. The first of these papers described a "disk-globe-and-cylinder" integrator invented by Professor James Thomson.[11] This invention was the spur that led Lord Kelvin to be the first to conceive of a general-purpose analogue device capable of automating the solution of second and higher-order linear differential equations.

In 1876 Lord Kelvin published an account of how two of the "disk-globe-and-cylinder" integrators invented by his brother could be cascaded together to solve second-order differential equations. He realised that the process could be automated by feeding back the output of the second integrator to the input of the first, and concluded that:

> the general differential equation of the second order with variable coefficients may be rigorously, continuously, and in a single process solved by a machine.[12]

In a subsequent paper Lord Kelvin described how this idea could be extended to solve higher than second-order differential equations.[13] In terms of the future development of general-purpose analogue computers, what Lord Kelvin had discovered was the significance of "closing the loop" through the use of feedback, and how this permitted the mechanisation of the solution of any order differential equation. Though Lord Kelvin laid the theoretical foundations for the development of what he referred to as a "differential analyzer", he never actually succeeded in building one. The chief technical reason for this was that the torque output from the second integrator was insufficient, without amplification, to drive the disk of the first integrator stage. Thus, it was the absence of a suitable torque amplifier that stopped Lord Kelvin from fully realising the first operational differential analyzer.

Network analyzers

In the early 20th century one other important source of impetus behind the development of analogue computing devices was the nature and increasing volume of computational problems associated with the rapid growth of electric power distribution networks. This area of electrical engineering design gave rise to the development of the first general-purpose mechanical analogue computer: the differential analyzer, built in the USA at MIT in the late 1920s. In the early 1920s the design of electric power generation and transmission systems also led to the development of a variety of electrical analogue computing devices and techniques, the foremost examples being the AC and DC network analyzers.

In the USA in the 1920s the General Electric Company and the Department of Electrical Engineering at MIT developed the first DC Network Analyzers.[14] DC network analyzers were originally developed for the purpose of calculating the short-circuit currents in electric power distribution systems. They were direct analogue devices consisting of a bank of resistance components which could be interconnected to construct a scaled electric analogue of the full-size distribution network. DC network analyzers were also employed in a variety of other applications including the steady-state analysis of fluid flow and temperature distribution studies.[15]

Notwithstanding their versatility in such supplementary applications, DC network analyzers were only of limited usefulness in the analysis of electric power distribution problems. Some of the shortcomings of DC network analyzers were overcome by the development of various alternating current devices known as AC calculating boards and/or AC network analyzers. The first large-scale AC network analyzer was constructed at the Department of Electrical Engineering at MIT in the USA in the late 1920s. The development and construction of the analyzer was undertaken jointly by staff at MIT and the General Electric Company.[16] In the 1920s and 1930s similar development work on a range of analyzers for electric power distribution studies was also being undertaken by the Westinghouse Electric Company.[17]

These AC network analyzers (or calculating boards) differed from the DC network analyzers in that the resistance components were replaced by a combination of resistance, capacitance and inductance components. This had the advantage that the phase relationships between the voltages and currents in different parts of the electric power distribution network could be represented directly. To keep costs down, passive components were generally used wherever possible. However, if a particular aspect of a system could not be satisfactorily represented by passive (i.e. resistance, capacitance, inductance) components, then transformers and amplifier circuits were also employed.

AC network analyzers, like the DC variety, proved to be extremely useful in applications other than the electrical network problems for

FIGURE 2.1 AC NETWORK ANALYZER[18]

which they were originally developed. The AC network analyzers were particularly useful in the solution of partial differential equations and mathematical problems characterised by many degrees of freedom. In addition to problems in electrical engineering AC network analyzers were used for a diverse range of applications, including problems in mechanical engineering, aircraft design and thermodynamics, (see Figure 2.1).[19]

The versatility of the network analyzers was predicated on the analogous nature of dynamic physical systems, by which the one physical phenomenon could be used to model and analyze the behaviour of another physical phenomenon. In the case of the AC network analyzer, the analogous nature of physical phenomena can be demonstrated by comparing the simple mechanical system shown below with its electrical analogy, see Figure 2.2. The mechanical system consists of a weight suspended by a spring with a friction brake for damping, see Figure 2.2a. The electrical system consists of a capacitor, inductor and resistor, see Figure 2.2b.

The response of the mechanical system can be described by the equation as shown. In the transposition from the mechanical systems to the electrical analogue electrical resistance (R) is equated to

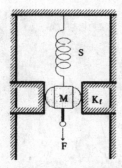

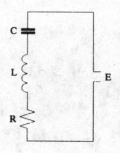

Fig 2.2 a. Mechanical system :
spring, mass and friction damping.

Fig 2.2 b. Electrical analogue :
capacitor, inductor and resistor.

Mechanical system: $F = M \dfrac{d^2 x}{dt^2} + K_f \dfrac{dx}{dt} + Sx$

$$F = M \dfrac{dv}{dt} + Kv + S \int v\, dt$$

Electrical system: $E = L \dfrac{d}{dt} + R + \dfrac{1}{C} \int 1\, dt$

$\Big\}$ where $M \simeq L$, $K \simeq R$, $S \simeq \dfrac{1}{C}$

FIGURE 2.2 THE ELECTRICAL ANALOGUE OF A MECHANICAL SYSTEM

friction (K_f), the reciprocal of electrical capacitance (C) is equated to the stiffness of the spring (S), and the electrical inductance (L) is equated to mass (M). Furthermore, the current that flows in the electric analogue represents the velocity (v) of the mechanical mass, and the electromagnetic force (E) corresponds to force (F) in the mechanical system. Here the primary aim is to study the oscillation of the mechanical system after a downward force (F) is briefly applied. What advantage then is there to be gained by employing electrical analogue techniques? The chief advantage is that it is generally easier to modify and study the effect of parameter variation in the electrical systems than in the mechanical. For example, in practice it is often much easier to devise and use a variable capacitor than to construct a spring with variable stiffness. The relatively low cost of electrical components, compared with mechanical parts, also favoured the use of electrical analogue techniques during the research and design phase.

By 1940 at least nine large-scale AC network analyzers had been built in the USA. In the USA and Europe the development and use of network analyzers continued after the end of World War II. By the mid-1950s the number of large AC network analyzers in operation in the USA had increased to more than thirty three. This increase corresponded to substantial growth in the installed power generation capacity and in the size of the distribution networks in the USA between 1945 and 1955.[20] However, when H. P. Peters carried out a market survey in 1954, he found that there was overcapacity in network analyzer equipment. Consequently, a number of institutions and firms were unable to maintain their installations because of the drop in demand from outside customers.[21]

Furthermore, by the late 1950s the importance and appeal of network analyzers as novel technologies for academic research and for teaching purposes had begun to wane.[22] The Department of Electrical Engineering at MIT, which had pioneered the development of large-scale AC network analyzers, sold their analyzer in the mid-1950s. Gordon S. Brown, Head of the Electrical Engineering Department at MIT gave the following reasons:

> We found that we could not afford it unless we rented it to power companies. Our gross revenue averaged about $10,000 per year. By the time we paid the salary of operating assistants, this revenue just about took care of expenses. Because of the need to schedule its use for power companies, its availability as an educational tool was often subordinated to second place. This was a deterrent to its uninhibited use by graduate student and staff research in the power area. ...There are now thirty-odd network analyzers in operation in the United States. Hence, we see no need to place high priority upon the education of electrical engineers in the mere use of an analyzer.[23]

The development of network analyzers started before, ran parallel to, and sometimes overlapped with the development of the general-purpose electronic analogue computers. To an extent they were competitive technologies, since many problems could be solved equally well using either of these two analogue computing technologies. Yet the overlap in application domains was by no means complete. Electronic analogue computers were not well suited to problems requiring the solution of partial differential equations. However, general-purpose electronic analogue computers had the advantage that they were more accurate/precise and flexible, though they were also more expensive because of the use of active, operational-amplifier-based computing elements. Though there was some convergence between the network analyzers and the postwar electronic analogue computers, they maintained distinct lineages, but nevertheless shared a similar conclusion; their displacement by electronic digital computers.

MECHANICAL ANALOGUE COMPUTERS IN THE USA AND BRITAIN

In the USA and Britain in the early 20th century, military applications provided much of the spur for the continuing development of wheel-and-disk and ball-and-disk integrators and other mechanical analogue computing devices. In particular, the development of Naval gun-fire controllers generated the bulk of the demand and posed the greatest technical difficulties. These gun-controllers/directors were special-purpose devices designed to solve a pre-determined set of equations, for a fixed number of variables, including the relative position, direction and speed of the target ship. In the USA the Ford Instrument Company was one of the leading firms in this area. In Britain much of the research and development work was undertaken by large engineering and armaments firms, such as the Metropolitan-Vickers Co.

Despite the pioneering work of the physicist Lord Kelvin in the late 19th century, little interest was shown in the development of generalised analogue computing devices until the 1920s. The principal figure responsible for the revival of interest in the construction of a "differential analyzer" was Vannevar Bush.

Vannevar Bush and differential analyzer development in the USA

In 1909 Vannevar Bush enrolled at Tufts College as an undergraduate. The intellectual environment which he entered was typical of early 20th century American engineering schools, which stressed the value of graphical representation and intuition rather than abstract rigour in mathematics and physics. In keeping with this tradition, the curriculum placed much of its emphasis on the attainment of laboratory and mechanical skills.[24]

Bush's interest in calculating machines began early in his career. In 1913, while studying for his master's degree, he designed, built and patented a surveying device he called a Profile Tracer. Bush began his doctoral studies at the Massachusetts Institute of Technology in 1915 and completed them in only one year. In 1919 Bush joined the staff of the Department of Electrical Engineering at MIT. Prior to his appointment to the faculty at MIT, Bush had spent several years teaching at Tufts College and a year working for General Electric.[25]

Many of the problems the electrical engineers at MIT were concerned with in the 1920s related to the emerging technologies of telephone communication, thermionic valves and the transmission of electrical power over long distances. Bush, who taught a course in electric power transmission at MIT, was especially aware of the computational difficulties which arose from this class of problem. The investigation of the stability of electrical transmission systems was often a laborious process involving the use of charts, tables of mathematical functions and the subsequent plotting by hand of the product of these functions. Further difficulties arose in the evaluation of the integral of the product of functions. In 1925 Bush reasoned that a machine might be capable of performing this operation and assigned Herbert R. Stewart, a graduate student, to work on the problem. Subsequently, in 1927, Bush published details of a device called a "continuous integraph", also referred to as the "product integraph".[26]

Bush's product integraph solved first-order differential equations directly and could also deal with second-order differential equations by a series of successive approximations. It used a modified watt-hour meter, of the type used to record domestic electricity consumption, to perform the integration of the product of two functions and a servo-driven pen recorder to plot the integral continuously. The product of a secondary function and the integral (of the product of two functions) was achieved mechanically. The functions to be integrated were first plotted on paper by hand. These were then entered by an operator who traced the line of the curves with a pointer as the calculation proceeded. The product integraph was applied to the solution of a variety of problems in advanced electrical transmission line theory and in bending moment calculations relating to the deflection of structural beams.

Although a powerful computational aid, it was in practice limited to the solution of first-order differential equations. Many of the problems which confronted engineers and scientists, however, required the solution of second and higher-order differential equations. Harold L. Hazen, a research assistant at MIT at this time, suggested to Bush adding a second level of integration to the product integraph to enable it to deal with more complex problems. The mechanical integrator which Hazen proposed was of the wheel-and-disc type. Bush is reported to have recognised immediately the fundamental significance of the step recommended by Hazen, and overnight Bush produced a twenty-page description of the generality of the innovation.[27]

Under the supervision of Bush, Hazen designed a revised version of the product integraph. The new machine retained the watt-hour meter integrator of the original design, to which a wheel-and-disk integrator was added. The two-integrator machine proved very useful and provided solutions with an accuracy of between one and three per cent.[28] However, it too had its shortcomings, which in part stemmed from the use of two different integrating devices. The modified product integraph led Bush to envisage the design of a new machine, based on the precisely machined mechanics of this simple yet mathematically powerful mechanical integrator. By 1928 he had secured funds from MIT for this project, and experimental work on the machine began.

In 1931 Bush published a paper entitled "The Differential Analyzer, A New Machine for Solving Differential Equations". This paper detailed the fruits of the labours begun in 1928.[29] The construction of the differential analyzer had in fact been completed in 1930. It had six wheel-and-disk integrators and was therefore capable of solving, for example, two third-order differential equations simultaneously. One of the problems encountered by Bush and Hazen during the design of the analyzer was the elimination of slippage between the wheel and the disk (of the integrator), while at the same time maximising the torque available from the disk shaft to drive other parts of the machine. This was one of the practical problems which Lord Kelvin had faced in realising his ideas for a differential analyzer in the late 19th century. However, since then C. W. Niemann had invented a practical method for torque amplification. Niemann invented the Torque Amplifier in 1927 while working at the Bethlehem Steel Company.[30] This device transformed precisely the rotation of an input shaft to an output shaft, amplifying the input torque by a factor of approximately 10,000. It was in fact Hazen who recommended the use of Niemann's torque amplifier to Bush, and it proved crucial to the success of the differential analyzer.[31]

The differential analyzer was a functional analogue computer; in other words, it was an equation solver which performed its calculations by modelling mathematical operations in mechanical forms. It operated on continuous data taken from curves and produced an output which was a continuous graphical representation of the solution. It was a long, narrow machine with four plotting tables connected to it on one side. Its six wheel-and-disk integrators were located in three protective glass-topped cases on the other side, see Figure 2.3. As a calculation proceeded, data were entered by operators who traced the outline of the input curves on the plotting tables. The output of the analyzer was a series of curves which were plotted automatically as the computation proceeded. It performed mechanically the operations of addition, subtraction, multiplication, division and integration.[32]

Almost half a century after Lord Kelvin had described how, in theory, a machine could be built that solved differential equations "rigorously, continuously, and in a single process", Bush and Hazen had succeeded in constructing such a machine. The differential analyzer was a great success and was praised for its flexibility and versatility. Although it was primarily applied to electrical engineering problems, it was also used in various scientific fields such as atomic physics.[33]

During the 1930s and early 1940s several differential analyzers similar to that developed at MIT were built in Britain, Germany, Norway and Russia. In the USA in 1935 the Moore School of Electrical Engineering of the University of Pennsylvania and the Ballistics Research Laboratory of the United States Ordnance Department in Aberdeen, Maryland, installed large-scale differential analyzers–built with the advice and under the supervision of Vannevar Bush.[34] In 1942 the General Electric Company installed America's fourth large-scale differential analyzer at its research laboratories in Schenectady.[35]

His successes in the design, construction and application of differential analyzers spurred Bush on to conceive of a more flexible machine. Bush recognised the limitations of the original design and realised that the manual setting-up procedure, which could

FIGURE 2.3 THE MIT DIFFERENTIAL ANALYZER[36]

sometimes take between two and three days, required speeding up.[37] In 1935 Bush and Samuel H. Caldwell embarked on the preliminary design of a new machine.[38] The following year the Rockefeller Foundation awarded MIT a sum of $85,000 and the project began in earnest. The machine, which came to be known as the "Rockefeller Differential Analyzer" or RDA, was intended to have thirty integrators and be able to run several programs simultaneously. In addition it was planned that the set-up procedure would be performed quickly and automatically via punched tapes. These design criteria were to be achieved by the use of electronics in many areas of the machine. In the mid-1930s the idea of using electronics in this application was a significant step forward. It is not surprising that recent advances in electronics were so quickly applied to differential analyzer design at MIT, as these machines were, after all, developed within MIT's Electrical Engineering Department.

The development of the RDA proved to be rather more difficult than Bush had anticipated: providing automatic programme control—the dynamic re-scheduling of computing elements—complicated the design enormously. As a result, the machine was not effectively operational until the middle of 1942 (three years behind schedule), and even in 1947 only eighteen integrators were functioning.[39] The statistics of the incomplete version in 1942 were nevertheless impressive: it weighed 100 tons, had 2,000 thermionic valves, 200 miles of wire and 150 motors.[40] The Rockefeller Differential Analyzer performed well and could be set up in only three to five minutes. The set-up data were, as planned, supplied automatically by means of three punched paper tapes: one contained the information on how the interconnections were to be made, the second set the values of the gear ratios, and the third set the values of the initial conditions. These paper tapes were stored so that, if required, a program could easily be repeated. The system also included automatic electrically controlled typewriters which were used to print tabulated results, see Figure 2.4.[41]

FIGURE 2.4 THE ROCKEFELLER DIFFERENTIAL ANALYZER[42]

Differential analyzer development in Britain

The first British differential analyzers were copied from the MIT machine of 1930. Yet the British context into which the American machine was transplanted differed markedly from its original setting. In the USA the differential analyzer was developed at an American engineering school where the solution of problems in electrical engineering provided the computational demand and principal spur for its construction. In Britain the leading proponents of differential analyzers were scientists working at universities in the fields of quantum mechanics, theoretical chemistry and mathematics. The first and most influential of these was Douglas R. Hartree.

Douglas R. Hartree and the Manchester machine

From 1929 to 1937 Douglas R. Hartree was Beyer Professor of Applied Mathematics at Manchester University. In the late 1920s Hartree became involved in the study of wave mechanics. One of the major contributions he made, was the introduction of the theory of Self-consistent Fields. This theory required that the computation of the wave function of electrons in an electromagnetic field be repeated many times. This in turn involved the solution of a great many second-order differential equations. In the late 1920s and early 1930s this was a laborious task which had to be carried out either by hand or on a desk calculator.[43]

The publication in 1931 of details of the development of the differential analyzer at MIT led Hartree to visit Boston in the summer of 1933.[44] Hartree's goal was to determine if the machine could be used to ease the burden of computation required in his work on self-consistent fields. Hartree was not disappointed and was soon using the differential analyzer to compute the wave function of Mercury.[45] Shortly after Hartree returned to Manchester he undertook to build a model of the MIT machine to expedite the computations required to complete his work on the atomic field of Mercury. Hartree enlisted one of his research students, Arthur Porter, to help him. Together they built a four-integrator differential analyzer largely from Meccano (a construction kit usually thought of as a children's toy). Completed in 1934, it was reported to give an accuracy of 2%.[46] The Meccano analyzer gave considerably less precise results than the full-scale differential analyzer which operated to an average precision of

FIGURE 2.5 MANCHESTER MECCANO DIFFERENTIAL ANALYZER[47]

0.1 percent. For many problems the relatively low precision was not a serious limitation, and the Meccano analyzer did much useful work, (see Figure 2.5).[48]

The combination of a desire for greater precision and the success of the model analyzer persuaded Hartree to seek funds for the construction of a differential analyzer on a scale similar to the one at MIT. The funds for the construction of a full-scale machine were donated by Robert McDougall, the deputy treasurer of Manchester University. McDougall was so impressed by the model machine that he provided funding for the project from his own personal resources. Work on the construction of the full-scale machine began in 1934. It was built for Manchester University by Metropolitan-Vickers Electrical Company Ltd., using drawings supplied by Bush. The Manchester machine was completed and installed in the basement of the Physics department at Manchester University in 1935, (see Figure 2.6).[49] It had four wheel-and-disk integrators and was therefore smaller than the MIT machine, which had six. However, shortly after its inauguration, McDougall made a second donation, bringing

FIGURE 2.6 THE MANCHESTER DIFFERENTIAL ANALYZER[50]

the total to £6000, and thus provided for the manufacture of four more wheel-and-disk integrators, enabling the final machine to solve eighth-order differential equations.[51]

In the meantime, however, developments in wave mechanics had led to a more complicated wave equation, the Hartree-Fock equation, the solution of which lay outside the scope of the differential analyzer. The Manchester machine was therefore never used for the purpose for which it was originally intended. The machine was instead used in a wide variety of applications and attracted researchers from government and industrial laboratories and contributed much to scientific computation.[52] Among those at Manchester University who became interested in the differential analyzer were the Professor of Physics at Manchester, P. M. S. Blackett, and F. C. Williams. Together they developed an automatic curve follower to obviate the practice of entering all mathematical functions and data manually.[53]

Hartree was promoted to Professor of Theoretical Physics at Manchester in 1937 and remained in this position until 1946. During World War II Hartree was seconded to the Ministry of Supply. The war brought with it a vast increase in the bulk and complexity of computational problems that required urgent solution. At Manchester the differential analyzer was requisitioned by the MoS for the war effort, and throughout it was in near-constant use. It was employed in a variety of problems including the study of anomalous radar pulses in the troposphere, calculations relating to the development of the radar cavity magnetron, the study of anti-aircraft gun controllers and the computation of the trajectories of projectiles used to provide the armed forces with ballistics tables.[54]

In the early 1940s the Manchester machine was not the only differential analyzer available for war work. The success of the Manchester projects had encouraged others to develop machines of their own. Among those who installed differential analyzers in the 1930s and early 1940s were the Physics Department of Queen's University Belfast, the Physics Department of the University of Birmingham, the War Office Projectile Development Establishment at Fort Halstead, the Coastal Artillery Experimental Station, the Armament Research Department at the Woolwich Arsenal, the General Electric Company Ltd, and the Valve Research Department of Standard Telephones and Cables Ltd. Yet these were all small-scale or model analyzers. The only other full-scale differential analyzer which was operational before the end of World War II was the machine built for Cambridge University by the Metropolitan-Vickers Company.[55]

Cambridge University and differential analyzer development
The pattern of differential analyzer development at Cambridge University was similar to that at Manchester. At both universities it was a leading scientist who was most active in promoting the development and use of differential analyzers. Also, the primary motivation in both cases was the construction of a machine to ease the burden of scientific computation they faced in their research.

In 1935 John Lennard-Jones, the Professor of Theoretical Chemistry at Cambridge University, supervised the design and construction of a Meccano differential analyzer. Lennard-Jones used the Manchester machine as the basis for the design and liaised with Hartree and Porter to make several improvements on the original.[56] Though the machine was primarily intended to serve Lennard-Jones's group of theoretical chemists, it was soon made available to other departments at the university. Unlike Hartree, however, Lennard-Jones never became involved in the day-to-day running of the Meccano analyzer. Instead, Lennard-Jones gave responsibility for the construction and operation of the machine to a colleague, J. B. Bratt.[57] In 1936, however, Bratt left the department, and

Lennard-Jones asked M. V. Wilkes, then a research student at the Cavendish Laboratory, to take responsibility for providing technical assistance to those wishing to use the machine. M. V. Wilkes had gained an understanding of the operation of the Meccano analyzer from using it to solve differential equations related to his study of the propagation of radio waves.[58]

Lennard-Jones saw that the model analyzer was a great success and persuaded the university to agree to fund the construction of a full-scale differential analyzer. However, the new machine was not intended to be used solely by the Department of Theoretical Chemistry, but was instead to be housed along with the desk calculators and several other computing devices to form a central computing facility known as the Mathematical Laboratory. Work on the analyzer began in 1937, and Wilkes was given the responsibility for overseeing its design and construction. The contract to build the differential analyzer was placed with Metropolitan-Vickers Ltd who had manufactured the Manchester machine. Unfortunately, the increase in the number of military contracts placed with Metropolitan-Vickers meant that they were unable to complete the Cambridge machine until late in 1939. It was therefore not installed at Cambridge until after the start of the war and was immediately requisitioned by the MoS for the duration of the war.[59]

A period of reappraisal: mechanical differential analyzers in Britain, 1945–1955

After the end of World War II, plans existed to continue to use the Manchester and Cambridge machines, and to work on the development of a new large differential analyzer for the National Physical Laboratory (NPL). At the same time, however, attention was increasingly drawn to the future of digital computing. Hartree and Wilkes were among the first in Britain to learn of the recent developments in digital computing in the USA. As their interest in digital computing grew, their involvement in analogue computer development rapidly waned.

In 1945 Hartree visited the USA to familiarise himself with the latest developments in digital computing. Hartree went to Harvard, where he saw the Automatic Sequence-Controlled Calculator, and to the Moore School of Electrical Engineering at the University of

Pennsylvania to see the as yet incomplete ENIAC. The ENIAC (Electronic Numerical Integrator and Computer) was built for the US Army's Ballistic Research Laboratory, Aberdeen, Maryland, primarily to calculate the trajectory data of projectiles. Approval for the construction of ENIAC was given by the US Army in 1943, but it was not completed until after the end of World War II. Consequently, in 1946 Hartree was invited to return to the USA by Colonel Paul Gillon to advise on the non-military use of the ENIAC.[60] Hartree took this opportunity (as he had during his visit to MIT in 1933) to use the new machine to perform calculations of his own and thus gained an understanding of how the machine operated.

Shortly after his return from the USA, Hartree left Manchester and moved to Cambridge University where he took up the Plummer Chair of Mathematical Physics. He left behind him not only his "Great Calculating Machine" ut also most of his involvement in analogue computing. Hartree almost certainly knew of the secret wartime developments in digital computing in Britain, including the Colossus code-breaking machine built at the British Code and Cipher School at Bletchley Park. Yet is seems that it was what he learned during his visits to the USA that convinced Hartree to pursue and promote digital instead of analogue computing. Thus, Hartree's inaugural lecture at Cambridge University in October 1946 concentrated on the American ENIAC computer. He gave details of its design, construction and method of operation and described his experience in using it to solve differential equations. Hartree now clearly believed that the future of automated computing lay in the pursuit of the digital approach, and at the start of the lecture he announced:

> ... it is time that physicists—and others—began looking ahead and thinking what they would like to do when (digital computing) equipment ... becomes more generally available than it is at the present.[61]

Moreover, earlier in 1946 Hartree had used his considerable influence to help the NPL secure funding from the Department of Scientific and Industrial Research (DSIR) for the construction of a digital computer. On 19 March 1946 the Executive Committee of the DSIR met to discuss the proposal to develop ACE (Automatic Computing Engine) for the Mathematical Division of the NPL.[62]

During the meeting Hartree bolstered the case in favour of the ACE project by arguing:

> ... if the ACE is not developed in this country the Americans will sweep the field. ... this country has shown much greater flexibility than the Americans in the use of mathematical hardware.[63]

and he continued:

> ... the machine should have every priority over even the existing proposal for the construction of a large differential analyzer (to be built at the NPL).[64]

As Andrew Hodges points out in his biography of Alan Turing, the mathematician who was largely responsible for the design of the ACE computer, Hartree's generous recommendation constituted "a remarkably smooth victory for the digital computer over the analogue approach."[65] Especially so when one considers the time and energy Hartree had spent in developing and using differential analyzers. The committee gave its approval for the development of a pilot version of the ACE computer. For the time being, funding was to be provided without sacrificing the existing differential analyzer project at the NPL. However, by 1949 plans to build a large differential analyzer at the NPL had been scrapped and instead a German firm was contracted to design and manufacture the NPL's new analogue machine.

The NPL became directly involved in the use and development of differential analyzers shortly after the Mathematical Division was formed in April 1945. The Superintendent of the division was John R. Womersley who in 1937 had co-authored a paper with Hartree on the application of the Manchester machine for the solution of certain types of partial differential equations.[66] The Division was initially organised into five sections: the Differential Analyzer, General Computing, Punched Card Machines, Statistics and ACE Section.

Initially, the Differential Analyzer section of the NPL consisted entirely of the staff and equipment associated with the Manchester machine. In 1946 Hartree handed over responsibility for the differential analyzer at Manchester to the Mathematics Division of the NPL. Yet the machine itself was not moved from Manchester to Teddington, where the NPL was located, until 1948. By this time the machine had been operating for over 13 years and was in need of extensive restoration. This and other factors, including the design and development of a new differential analyzer at the NPL, meant that little work was performed on it prior to its replacement in 1954.[67]

The task of designing the new 24-integrator differential analyzer took up much of the time of J. G. L. Michel, the section head, and other senior staff of the Division. As a result, they were not able to provide sufficient support to users of the existing differential analyzer or to undertake very much service computing. To ease the work load, the NPL's own project to design a differential analyzer was cancelled. Instead, in March 1949 an agreement was reached with the Admiralty for the NPL to take over an existing contract for a new differential analyzer which the Admiralty had placed with the German firm of Schoppe & Faeser. Completed in 1954, the NPL machine was the largest and last of its kind to be installed in the UK. It had 20 high-precision mechanical integrators, yet it also incorporated electronic switching and was programmed from a central plug board which enabled it to run three different problems simultaneously.[68] Nevertheless, a great deal had changed since the contract for the differential analyzer was first approved, and by 1954 research in numerical analysis and electronic digital computing overshadowed the differential analyzer work at the NPL. The analogue machine remained in operation only until 1958, when its use was discontinued.[69]

After the war the Mathematical Laboratory at Cambridge University underwent considerable reorganisation, and the development of digital computing took highest priority. In August 1945 Lennard-Jones resigned as director of the Mathematical Laboratory and took up the post of Director General of Scientific Research (Defence) for the Ministry of Supply.[70] M. V. Wilkes was confirmed as the new Director of the Mathematical Laboratory in October 1946. Wilkes had been working as acting director since his return to Cambridge in September 1945 from war service with the RAF.[71] He had already started to formulate plans for the reorganisation of the laboratory once the Ministry of Supply relinquished control over the facility. The direction taken by Wilkes was greatly influenced by the news of developments in the USA brought back by Hartree and

L. J. Comrie, the director of Scientific Computing Service, Ltd (SCS). In 1946 Comrie lent Wilkes a copy of the *First Draft of a Report of the EDVAC* (Electronic Discrete Variable Automatic Computer), in which John von Neumann described the logical operation and possibilities of the electronic digital computer.[72] Subsequently, with the assistance of Hartree, Wilkes visited the USA and attended the now famous lectures on the design of digital computers at the Moore School of Electrical Engineering. When Wilkes returned to Britain he was determined to pursue the development of a machine based on the EDVAC proposal.[73] By 1947 Wilkes had established a new set of priorities for the Mathematical Laboratory which placed less emphasis on the provision of a computing facility for staff and the running of courses in computing, and more on research into the design of new computing machinery. In fact, the main goal of the laboratory had become the design and construction of an electronic digital computer known as the EDSAC (Electronic Delay Storage Automatic Calculator).[74]

Thus by 1950 the Manchester machine had been dismantled and removed to the NPL, and at Cambridge Hartree and Wilkes set their interest in analogue computing to one side and instead pursued and promoted the digital approach to automated computation. Postwar analogue computing was not a continuation of the work at the leading prewar centres. Nevertheless, the development and construction of new mechanical differential analyzers did not come to a sudden stop. Several new machines were built in Britain between 1945 and 1955. This work, however, was not undertaken in academia but in government research establishments and in industry. Table 2.2 lists the principal mechanical and electro-mechanical differential analyzers in operation in Britain between 1953 and 1956.

SUPERSEDING BUT NOT REPLACING: MECHANICAL DIFFERENTIAL ANALYZERS AND ELECTRONIC ANALOGUE COMPUTERS

Though commercial general-purpose electronic analogue computers were available in the USA by the late 1940s, and in Britain by 1953, the replacement of the mechanical differential by the electronic analogue computer was neither unproblematic nor swift. The principal appeal of the electronic analogue computer, *vis-à-vis* the mechanical differential analyzer, was its higher computing speeds, lower costs

TABLE 2.2 DIFFERENTIAL ANALYZERS INSTALLED IN BRITAIN, 1953–1956[75]

Location	No. of Integrators	Type
Building Research Establishment, (DSIR), Garston, Watford, Herts.	10	Mechanical (under construction in 1953)
Courtaulds Research Laboratories, Maidenhead, Berks.	8	Mechanical
ICI, Butterworth Research Laboratories, Welwyn, Herts.	6	Mechanical
Royal Military College of Science, Shrivenham, Wilts.	8	Mechanical
	4	Mechanical/Meccano
Army Operation Research Group, Byfleet, Surrey	6	Mechanical
National Physical Laboratory, Teddington, Middlesex	8	Mechanical (Former Manchester machine. Removed in 1954)
	20	Electro-mechanical (Installed 1954)
Radar Research Establishment, Great Malvern, Worcs.	4	Electro-mechanical
Royal Aircraft Establishment, Farnborough, Hants.	6	Electro-mechanical (Replaced by DEUCE digital computer)
Elliott Bros, Research Laboratories, Borehamwood, Herts.	6	Electro-mechanical

and the relative simplicity of its construction and operation: wires rather than shafts and gear trains connected computing components, and operational amplifiers were smaller and lighter than wheel-and-disk integrators. However, the mechanical machines had two significant advantages: greater accuracy and greater mathematical flexibility.

Comparing mechanical and electronic systems, we find that postwar electro-mechanical differential analyzers could operate to an overall accuracy of 0.1%. Until the mid-1950s most electronic analogue computers operated to an accuracy of 1%. The relatively low accuracy/precision of electronic analogue computers improved as new operational amplifiers and components with tighter tolerances were developed, and as component layout, computer wiring and overall construction improved. The source of the greater mathematical flexibility of the mechanical differential analyzers was that integration could be performed with respect to any variable, dependent or independent. DC operational amplifiers were in general restricted to integration and differentiation with respect to a single independent variable, time.[76] As a consequence, electronic analogue computers were only partially appropriate, and thus successful, as replacements for the mechanical differential analyzers.

Nevertheless there were a great many problems, particularly in engineering, for which neither the relatively low precision of the electronic analogue computer nor its mathematical limitations diminished its usefulness. Indeed, the specific characteristics of electronic analogue computers made them particularly well suited to the study of dynamic physical systems in which time is the independent variable. DC operational amplifier-based analogue computers can be set up to model these dynamic systems and can then be used to study them through a process of simulation. With time as the independent variable, the time scale of the simulation can be varied so that problems can be solved in either compressed, real-time or extended time scales.

The ability to use a compressed time scale was an especially useful feature of electronic analogue computing, as it permitted a large number of solutions/simulations to be made for a range of parameters in a relatively short time. Real-time operation was a particularly advantageous feature of electronic time integration/differentiation, because this permitted "hardware in the loop" studies. This meant that parts of the dynamic system being designed could be included in the simulation model. Often the "hardware" consisted of experimental components for which a full mathematical definition was being sought and non-linear components that were not amenable to mathematical analysis.[77]

The ability of the electronic analogue computers to compute rapidly and the relative ease with which they could be used for problems in dynamics meant that they were particularly well suited to the study of aeronautics and guided weapons, the salient characteristics of these fields of study being the use of trial-and-error and parameter-searching design methods, and the practice of performing a large number of computations to investigate the influence of random effects created by internal and external "noise", and/or by pilots. Though these characteristics are not unique to the development of aircraft and guided weapons, alternative design methods involving direct experimentation and field trials are especially expensive and hazardous.

Indeed, the relationship between aeronautics/guided weapons and analogue computing is such that it is difficult to overstate the influence of these fields of study on the development of the electronic analogue computers. As we shall see, several of the commercial electronic analogue computer systems originated in programs to develop military aircraft and guided weapons.

NOTES

1. In other words, whereas in digital machines quantities are converted into discrete numerical values, in analogue computing devices quantities are converted into continuously variable magnitudes which are represented, for example, by the rotation of a disk or shaft, or by the value of a voltage or current.

2. S. Fifer, *Analogue Computation: Theory, Techniques and Applications*, 4 Vols., McGraw-Hill, New York, NY, 1961, Vol. 1, p. 4.

3. Though they were versatile, the fact that the network analyzers were not functional analogue devices limited the scope of their operation and they were generally considered multipurpose rather than general-purpose analogue computers.

4. A. B. Clymer, "The Mechanical Analogue Computers of Hannibal Ford and William Newell", *Annals of the History of Computing*, Vol. 15, no. 2, 1993, pp. 19–34.

5. The concept of a general-purpose analogue computing device has been the source of some controversy, since some have argued that by comparison with the digital computer they are all inherently special-purpose (see chapter 8).

6. M. Williams, *A History of Computing Technology*, Prentice-Hall, Englewood Cliffs, NJ, 1986, pp. 111–118.

7. F. J. Murray, *Mathematical Machines, Volume II: Analogue Devices*, Columbia University Press, New York, NY, 1961; *New Encyclopedia Britannica*, University of Chicago, Chicago, Vol. 9, 1991, pp. 496–497.

8. G. Kirchhoff, *Annals of Physics, Leipzig*, Vol. 64, 1845, p. 497; G. Kirchhoff *Collected Papers*, Leipzig, 1882, p. 56; C. F. Kayan, "An Electrical Geometrical Analogue for Complex Heat Flow", *Trans., ASME*, Vol. 67, 1945, pp. 713–718.

9. Three dimensional analysis was also possible, though more complicated and less often used. S. Softky and J. Jungerman, "Electrolytic tank measurements in three dimensions", *Review Scientific Instruments*, Vol. 23, 1952, pp. 306–307.

10. W. G. Adams, *Proceedings Royal Society*, Vol. 24, 1876, p. 1; A. R. Boothroyd, E. C. Cherry and R. Makar, "An electrolytic tank for the measurement of steady-state response, transient response, and allied properties of networks", *Proc., IEE*, Vol. 96, pt 1, 1949, pp. 163–177.

11. This planimeter was never built commercially, but his brother Sir William Thomson—later Lord Kelvin—incorporated the integrator into a device which he called a Tidal Harmonic Analyzer, used to predict the height of tides. J. Thomson, "On an Integrating Machine

Having a New Kinematic Principle", *Proc., Royal Society*, Vol. 24, 1876, pp. 262–265; W. Thomson, "On an Instrument for Calculating $\int \phi(x)\tau(x)\,dx$, the Integral of the Product of Two Functions", *Proc., Royal Society*, Vol. 24, 1876, pp. 266–268; W. Thomson, "Harmonic Analyzer", *Proc., Royal Society*, Vol. 27, 1878, pp. 371–373.

12. W. Thomson, "Mechanical Integration of Linear Differential Equations of Second Order with Variable Coefficients", *Proc., Royal Society*, Vol. 24, 1876, pp. 269–271.

13. W. Thomson, "Mechanical Integration of the General Linear Differential Equation of Any Order with Variable Coefficients", *Proc., Royal Society*, Vol. 24, 1876, pp. 271–275.

14. The first of several DC Network Analyzers developed at General Electric, details of which were first published in 1920, remained in operation for more than 25 years. H. A. Peterson and C. Concordia, "Analyzers for Use in Engineering and Scientific Problems", *General Electric Review*, Vol. 49, 1945, pp. 29–37; K. L. Wildes and N. A. Lindgren, *A Century of Electrical Engineering and Computer Science at MIT, 1882–1982*, MIT Press, Cambridge, Mass., 1985.

15. For example, a temperature distribution study could be conducted by the construction of a network of electrical resistances that models the distribution of elements that resist heat flow in the real system. Measurements of voltages at points in the lattice of resistance components could then be interpreted as measurements of temperature distribution.

16. K. L. Wildes and N. A. Lindgren, *A Century of Electrical Engineering and Computer Science at MIT, 1882–1982*; H. A. Peterson and C. Concordia, "Analyzers for Use in Engineering and Scientific Problems", *General Electric Review*.

17. W. Aspray, "Edwin L. Harder and the Anacom: Analogue Computing at Westinghouse", *Annals of the History of Computing*, Vol. 15, no. 2, 1993, pp. 35–52; E. L. Harder and G. D. Cann, "A Large-Scale General-Purpose Electric Analogue Computer", *Trans., AIEE*, Vol. 67, 1948, pp. 664–673.

18. H. A. Peterson and C. Concordia, "Analyzers for Use in Engineering and Scientific Problems", p. 3.

19. Ibid., pp. 7–8.

20. For example, installed capacity grew from 52,000 megawatts in 1947 to 91,000 megawatts at the end of 1953. H. P. Peters, "AC Network Calculator Market Survey", *Typescript report*, Georgia Institute of Technology, Atlanta, Georgia, 1954, pp. 1–8.

21. Ibid.

22. K. L. Wildes and N. A. Lindgren, *A Century of Electrical Engineering and Computer Science at MIT, 1882–1982*, pp. 103–104.

23. G. S. Brown, quoted in, H. P. Peters, "AC Network Calculator Market Survey", pp. 5–6.

24. L. Owens, "Vannevar Bush and the Differential Analyzer: The Text and Context of an Early Computer", *Technology and Culture*, Vol. 27, no. 1, 1986, pp. 63–95; E. T. Layton, Jr., *The Revolt of the Engineers: Social Responsibility and the American Engineering Profession*, Johns Hopkins University Press, Baltimore, 1986, pp. 53–79.

25. This device, detailed in his master's thesis—"An Automatic Instrument for Recording Terrestrial Profiles" consisted of an instrumentation box supported at both the front and rear by a bicycle wheel. By pushing the tracer over the ground, the instruments inside would record the profile (or undulation) of the land beneath it. This invention was never a commercial success, despite Bush's attempts to sell it to manufacturers. Bush's doctoral thesis was concerned with Oliver Heaviside's "operational calculus", in which differential equations were reduced to algebraic equations, making their solution considerably easier. K. L. Wildes and N. A. Lindgren, *A Century of Electrical Engineering and Computer Science at MIT, 1882–1982*, pp. 82–95; V. Bush, *Pieces of the Action*, William Morrow, New York, 1970, pp. 155–157; O. Heaviside, *Electrical Papers* Macmillan, New York, 1892.

26. V. Bush, F. D. Gage and H. R. Stewart "A Continuous Integraph," *Journal of the Franklin Institute*, Vol. 203, 1927, pp. 63–84. Gaspard de Coriolis is generally credited as being the

first to set out the fundamental principles of the integraphs in a publication in 1836. The integraph was a device to plot continuously the integral of curve which is being traced. By the late nineteenth century the first practical instruments were being built and refinements made. The developments in integraph design made by Abdank-Abakanowicz in France in the 1870s have been acknowledged as amongst the most noteworthy. Abdank-Abakanowicz later collaborated with the Swiss instrument maker Coradi to produce a series of commercial integraph devices. Yet little further development took place till the work by Bush and his colleagues at MIT in the 1920s. G. de Coriolis, *Journal de Liouville*, Vol. 1, 1836; Abdank-Abakanowicz, *Die Intergraphen*, B. G. Teubner, Leipzig, 1889; F. J. Murray, *Mathematical Machines, Volume II: Analogue Devices*

27. K. L. Wildes and N. A. Lindgren, *A Century of Electrical Engineering and Computer Science at MIT, 1882–1982*, p. 90.

28. V. Bush and H. Hazen, "Integraph Solutions of Differential Equations", *Journal of the Franklin Institute*, No. 204, 1927, pp. 575–615.

29. The term "differential analyzer" was first coined in 1876 by the British physicist Lord Kelvin in a paper to the Royal Society. In his memoirs, Bush claims no knowledge of this work prior to completion of his differential analyzer. V. Bush, *Pieces of the Action*, 1970; V. Bush and H. Hazen, "The Differential Analyzer, A New Machine for Solving Differential Equations", *Journal of the Franklin Institute*, No. 212, 1931, pp. 447–488.

30. C. W. Niemann, "Bethlehem Torque Amplifier", *American Machinist*, Vol. 66, 1927, pp. 895–897.

31. Hazen was also responsible for another notable feature of the differential analyzer: the "frontlash unit" which offset backlash effects in the gears and shafts.

32. In 1936 Bush described how the feedback or "back-coupling" of the wheel-and-disk integrators operated and how they could be combined to solve problems:

> A single integrator is merely an instrument for performing the operation of integration. If its disc be turned at constant speed, having at any time an angle x, and if its displacement be varied in accordance with a variable y, its output will yield the value of $\int y\,dx$. Suppose now, however, that its output be connected to the y shaft. This is back-coupling. It forces y at all instants to be equal to the output of $\int y\,dx$. The machine is now constrained, and can move only in one definite way, namely in accordance with the equation $y = \int y\,dx$ or $dy/dx = y$. It thus yields the exponential solution of this equation, if simultaneous readings be taken of the positions of the x and y shafts.
>
> ...The way in which an equation is placed upon the machine may be traced. It is first solved formally for the highest order derivative, and integrated once formally, to yield the derivative of next lower order. A shaft is assigned to this, and one to the independent variable. An integrator may now be connected to yield the derivative of next lower order, and so on until the dependent variable is reached. As we proceed in this manner, every term in the equation becomes represented by the revolution of a shaft. These shafts are then interconnected through differential gears so that their revolutions sum to zero. This closes the equation, and in effect provides for the drive of the shaft assigned to the next to highest order derivative. When the independent variable shaft is now turned, every other shaft is driven, and the machine is constrained to move in accordance with the equation.

V. Bush, "Instrumental Analysis", *Bulletin of the American Mathematical Society*, Vol. 42, 1936, pp. 649–669. pp. 663–664.

33. J. Crank, *The Differential Analyzer*, Longmans, London, 1947, p. 3.

34. In the early 1930s staff from the Moore School of Electrical Engineering and the Ballistics Research Laboratory (BRL) approached Bush with a view to obtaining a copy of the new machine. By 1933 a single design had been agreed upon to satisfy their different requirements. The BRL machine was built on a commercial basis under the supervision of Bush and his col-

leagues. Construction of the Moore School machine was funded by the Civil Works Administration, one of the agencies established by President Roosevelt to stimulate the economy during the Great Depression of the 1930s. H. H. Goldstine, *The Computer: from Pascal to von Neumann*, Princeton University Press, Princeton, 1972, p. 96; I. Travis, "Differential Analyzer Eliminates Brain Fag", *Machine Design*, Vol. 7, no. 7, 1935, pp. 15–18.

35. H. P. Kuehni and H. A. Peterson, "A New Differential Analyzer", *Electrical Engineering*, Vol. 63, 1944, pp. 221–228.

36. K. L. Wildes and N. A. Lindgren, *A Century of Electrical Engineering and Computer Science at MIT, 1882–1982*, p. 91.

37. J. Crank, *The Differential Analyzer*, p. 115.

38. Bush's administrative duties as MIT's vice-president and dean of engineering meant that it was necessary for Caldwell to bear much of the responsibility for the project. Furthermore, prior to the end of the Rockefeller Analyzer project Bush left MIT to become president of the Carnegie Institute of Washington. Shortly afterwards, on 27 June 1940, the Council on National Defence established the National Defense Research Committee (NDRC). Bush was appointed chairman, and MIT President Karl T. Compton was assigned responsibility for the newly formed Division D which dealt with gun fire control, radar, detection, and countermeasures. In June 1941 the Office of Scientific Research and Development (OSRD) was established and Bush was appointed director. The OSRD had two principal divisions: the Committee on Medical Research (CMR) and the NDRC. Prompted by the Japanese attack on Pearl Harbor on 7 December 1941, reorganization of the NDRC took place in 1942. Harold Hazen, then head of MIT's Electrical Engineering Department, was assigned control of the newly formed Division 7. This division was responsible for the development of gun fire control systems, for example, anti-aircraft gun directors. Development work which took place under the auspices of Division D and later Division 7 had a significant impact on the future of the analogue computer. However, other important developments in analogue computing had been taking place in America outside of MIT prior to World War Two. One of the leading figures in this field was George A. Philbrick

39. Because of wartime security measures information on the RDA was not released until 1945. V. Bush and S. Caldwell, "A New Type of Differential Analyzer", *Journal of the Franklin Institute*, Vol. 240, 1945, pp. 255–326.

40. K. L. Wildes and N. A. Lindgren, *A Century of Electrical Engineering and Computer Science at MIT, 1882–1982*, p. 92.

41. J. Crank, *The Differential Analyzer*, pp. 115–116; H. E. Rose, "The Mechanical Differential Analyzer: Its Principles, Development, and Applications", *Proc., Institute of Mechanical Engineers*, Vol. 159, 1948, pp. 46–54.

42. H. E. Rose, "The Mechanical Differential Analyzer: Its Principles, Development, and Applications", p. 54.

43. P. A. Medwick, "Douglas Hartree and Early Computation in Quantum Mechanics", *Annals of The History of Computing*, Vol. 10, no. 2, 1988, pp. 105–111; D. R. Hartree, "The ENIAC: An Electronic Computing Machine", *Nature*, Vol. 158, 1946, pp. 500–506.

44. V. Bush and H. Hazen, "The Differential Analyzer: A New Machine for Solving Differential Equations", pp. 447–488.

45. D. R. Hartree, "Approximate Wave Functions and Atomic Field of Mercury", *Physical Review*, Vol. 46, 1934, pp. 738–743.

46. D. R. Hartree and A. Porter, "The Construction and Operation of a Model Differential Analyzer", *Memoirs Manchester Lit. and Phil. Society*, Vol. 79, no. 5, 1935, pp. 51–73; H. H. Goldstine, *The Computer from Pascal to von Neumann*, Princeton University Press, Princeton, NJ, 1972, p. 95.

47. H. E. Rose, "The Mechanical Differential Analyzer: Its Principles, Development, and Applications", p. 54.

48. J. Crank, *The Differential Analyzer.*

49. D. R. Hartree, "The Differential Analyzer", *Nature*, Vol. 135, 1935, pp. 940–943.

50. H. E. Rose, "The Mechanical Differential Analyzer: Its Principles, Development, and Applications", p. 53.

51. C. G. Darwin, "Douglas Rayner Hartree: 1897–1958", *Biographical Memoirs of Fellows of the Royal Society*, Vol. 4, 1958, pp. 103–116, p. 107; Anon., "Differential Analyzer at Manchester University", *Engineering*, Vol. 140, July 26, 1935, pp. 88–90; Anon., "Differential Analyzer", *The Engineer*, Vol. 160, July 26, 1935, pp. 82–84.

52. C. G. Darwin, "Douglas Rayner Hartree: 1897–1958", p. 108; D. R. Hartree, "A Great Calculating Machine", *Proc., Royal Institution*, Vol. 31, 1940, pp. 151–170; D. R. Hartree, "The Bush Differential Analyzer and its Applications", *Nature*, Vol. 146, 7 Sept 1940, pp. 319–323; D. R. Hartree, "The Mechanical Integration of Differential Equations", *Mathematical Gazette*, Vol. 22, 1938, pp. 342–364; D. R. Hartree and A. K. Nuttall, "The Differential Analyzer and its Applications in Electrical Engineering", *Journal of the Institute of Electrical Engineers*, Vol. 83, 1938, p. 648.

53. P. M. S. Blackett and F. C. Williams, "An Automatic Curve Follower for Use With the Differential Analyzer", *Proc., Cambridge Philosophical Society*, Vol. 35, 1939, pp. 494–505.

54. M. V. Wilkes, in D. R. Hartree, *Calculating Instruments and Machines*, The Charles Babbage Institute Reprint Series for the History of Computing, Vol. 6, Tomash Publishers, 1984, p. xi; and C. G. Darwin, "Douglas Rayner Hartree: 1897–1958", p. 107.

55. J. Crank, *The Differential Analyzer*; H. S. W. Massey, J, Wylie, R. A. Buckingham and R. Sullivan, "Small Scale Differential Analyzer: its Construction and Operation", *Proc., Royal Irish Academy*, Pt. A, Vol. 45, 1938, pp. 1–21; R. E. Beard, "The Construction of a Small-Scale Differential Analyzer and its Application to the Calculation of Actuarial Functions", *Journal of the Institute of Actuaries*, Vol. 71, 1941, pp. 193–227; M. Croarken, *The Centralisation of Scientific Computation in Britain: 1925–1955*, PhD thesis, University of Warwick, 1985, pp. 84–85.

56. N. F. Mott, "John Lennard-Jones", *Biographical Memoirs of Fellows of the Royal Society*, Vol. 1, 1955, pp. 175–184.

57. J. B. Bratt, J. E. Lennard-Jones and M. V. Wilkes, "The Design of a Small Differential Analyzer", *Proc., Cambridge Philosophical Society*, Vol. 35, 1939, pp. 485–493.

58. M. V. Wilkes, *Memoirs of a Computing Pioneer*, 1985; M. V. Wilkes, "Automatic Computing", *Nature*, Vol. 166, 1950, pp. 942–943.

59. M. Croarken, *The Centralisation of Scientific Computation in Britain: 1925–1955*, pp. 79–84.

60. M. V. Wilkes, in D. R. Hartree, *Calculating Instruments and Machines*, 1984, pp. xi-xii

61. D. R. Hartree, *Calculating Machines: Recent & Prospective Developments and their impact on Mathematical Physics*, Cambridge University Press, Cambridge, 1947, pp. 6–7.

62. A. Hodges, *Alan Turing: The Enigma of Intelligence*, Burnett Books and Hutchinson, London, 1983, pp. 333–335.

63. Ibid., p. 335.

64. Ibid., p. 335.

65. Ibid., p. 335.

66. D. R. Hartree and J. R. Womersley, "A Method for the Numerical or Mechanical Solution of Certain Types of Partial Differential Equations", *Proc., Royal Society*, Pt. A, Vol. 161, 1937, pp. 353–366.

67. M. Croarken, *The Centralisation of Scientific Computation in Britain: 1925–1955*, p. 144.

68. J. G. L. Michel, "The Mechanical Differential Analyzer: Recent Developments and Applications", *Journees International du Calcul Analogique*, Sept 1955. The machine was not designed entirely by the German firm; Metropolitan-Vickers Co. Ltd was responsible

for the design of special amplifiers for the control of the servo-mechanisms. Two British firms were also involved in the task of installing the analyzer; the Plessy Co., of Ilford and the Strand Electric and Engineering Co. of London, were responsible for installing the electrical wiring to interconnect the separate units and control panel. See, Anon., "Differential Analyzer: The N.P.L.'s New Analogue Computer for Solving Differential Equations", *Engineering*, Vol. 178, Nov 19, 1954, pp. 659–660.

69. M. Croarken, *The Centralisation of Scientific Computation in Britain: 1925–1955*, p. 162.

70. During the war Lennard-Jones had worked for the MoS; in 1942 he was appointed Chief Superintendent of Armaments Research and left Cambridge. N. F. Mott, "John Lennard-Jones", *Biographical Memoirs of Fellows of the Royal Society*.

71. During the war Wilkes gained experience in electronics while working on radar and related subjects for the RAF. M. V. Wilkes, *Memoirs of a Computer Pioneer*, MIT Press, Cambridge, Mass., 1985, p. 34.

72. J. von Neumann, *First Draft of a Report on the EDVAC*, Moore School of Electrical Engineering, University of Pennsylvania, 30 June 1945.

73. M. V. Wilkes, *Memoirs of a Computing Pioneer*, p. 109.

74. M. Croarken, *The Centralisation of Scientific Computation in Britain: 1925–1955*, pp. 199–207.

75. A. J. Knight, "A Survey of Computing Facilities in the U.K. (2nd ed.)", *Directorate of Weapons Research Report No. 5/56*, Ministry of Supply, London, August 1956; C. A. Reiners, "Survey of Computing Facilities in the U.K.", *Directorate of Weapons Research (Defence) Report No. 13/53*, Ministry of Supply, London, Sept 1956.

76. Having time as the sole independent variable excluded the construction of direct analogues of partial differential though other techniques were developed for dealing with this type of equation on DC electronic analogue computers.

77. One of the most difficult parts of any man-machine system to model mathematically is the operator, driver or pilot. Real-time simulation allowed this person to be included in the simulation.

CHAPTER 3

The Origins, Form and Function of Electronic Analogue Devices and Computers, 1937–1950

INTRODUCTION

In the previous chapter we have seen that the foremost and most versatile of the analogue computing devices developed up to the late 1930s were the mechanical differential analyzers and the electrical network analyzers. Yet the history of these devices is not a prehistory of the electronic analogue computer. It is important to realise that the post-World War II electronic analogue computer owes little to either of these distinct paths in the development of analogue computing devices. Nor, though this is perhaps less surprising, is the electronic analogue computer related to developments in punched-card equipment and electronic digital computers. Neither the key technology—the DC operational amplifier—nor the general-purpose electronic analogue computer originated from attempts to progressively apply electronics to existing analogue or digital aids to computation in engineering and science. They were a largely autonomous development, originating from research and development in electronic devices for fixed-purpose control applications, and for the simulation and analysis of control systems, aircraft and guided missiles.

The first half of this chapter is concerned with the prewar and wartime origins of electronic analogue devices and computing systems. Here we see that in the late 1930s and early 1940s pioneering work on the use of thermionic valve amplifiers and alternating current signal techniques to model and simulate dynamic systems was undertaken independently by G. A. Philbrick in the USA and H. Hoelzer in Germany. The dynamic systems being studied by Philbrick in the late 1930s were industrial process controllers. In wartime Germany, Hoelzer was a member of staff at the Peenemunde centre for rocket development, where he worked on the development of guidance systems and the simulation of missiles. In wartime Britain, at the government-funded Telecommunications Research Establishment, Malvern, work on electronic analogue computing devices was carried out in connection with the development of anti-aircraft gun controllers, radar, aircraft and simulators. In the USA during the war, seminal work on the development of alternating and direct current amplifiers was carried out at the Bell Telephone Laboratories. At BTL these devices were developed for use in anti-aircraft gun controllers. Finally, we see that in the USA the first concrete steps to create a general-purpose electronic analogue computer were undertaken at Columbia University during the war.

The second part of the chapter introduces and describes the chief characteristics of postwar general-purpose electronic analogue computers. Here I indicate how DC operational amplifiers can be configured to perform computation and simulation, and give an overview of the major functional components of postwar electronic analogue computers.

FROM CONTROL TO COMPUTATION: ORIGINS OF THE ELECTRONIC ANALOGUE ALTERNATIVE

The origins of general-purpose electronic analogue computers lie in more than two decades of pre-World War II research and development in thermionic valves and signal amplifiers for radio and telecommunications. Two of the most significant contributions in thermionic valve and amplifier design in the inter-war years were made by H. Nyquist and H. S. Black while working at the Bell Telephone Laboratories in the USA. In particular, the invention of the feedback-stabilised amplifier by H. S. Black in 1934 laid much of the foundations for the subsequent development of DC operational amplifiers—the key technology in postwar electronic analogue computing. DC operational amplifiers are direct-current coupled amplifiers that can be configured to perform (or operationalise) the mathematical functions of integration, differentiation, addition and subtraction. Yet though the DC operational amplifier was the dominant postwar technology, the enrolment and redefinition of AC amplifiers and techniques for analogue computing, simulation and control applications preceded that of DC amplifiers. Until the late 1940s technical merits favoured neither AC nor DC amplifiers, and which, if either, would dominate was indeterminate.[1]

George A. Philbrick and Polyphemus

The first explicit use of AC amplifiers in an electronic analogue computing and simulation system was in the USA by George A. Philbrick. After graduating from Harvard's engineering school in communiations engineering in 1935, Philbrick went to work for the Atlantic Precision Instrument Company, a subsidiary of the Foxboro Instrument Co. At Foxboro, Philbrick initially spent much of his time on the mathematical analysis of process-control systems. Realising the problems that engineers faced in understanding increasingly complex and mathematically oriented control theory, he designed a special-purpose electronic analogue computer which he called the Automatic Control Analyzer (ACA). Built between 1938 and 1940, the analyzer was designed to simplify the study of control loops and to help engineers solve problems in the design of industrial process controllers.[2]

The ACA contained several thermionic valves and a number of passive electrical components that formed a fixed electronic model of a process controller in an industrial process. Parameters could be varied using controls on the front panel, and the simulated effects could be observed. These simulations were performed at high speed by compressing the time scale, and then repeated automatically at one of a number of repetition rates. The output data from the ACA was displayed on a single oscilloscope. This was the only form of data output device, and Philbrick nicknamed the ACA Polyphemus, after the one-eyed Cyclops who, according to Greek mythology, was blinded by Odysseus.

Polyphemus was a significant development in the history of electronic analogue computation for two reasons. Firstly, because Philbrick used electronic circuits to operationalise mathematical functions and model dynamic systems for the purpose of analysis and design. Secondly, with Polyphemus Philbrick founded a distinct tradition in the history of electronic analogue computing known as "repetitive operation". In 1942 Philbrick left the Foxboro Company to become Samuel Caldwell's technical aide at the National Defense Research Committee (NDRC). Caldwell, formerly in charge of the Rockefeller differential analyzer project at MIT, was at that time chief of Section 7.2 of the NDRC which was concerned with problems in airborne gun fire control. After the end of World War II, Philbrick founded his own company and developed and manufactured a range of repetitive-operation electronic analogue computing devices and systems.[3]

Helmut Hoelzer: electronic analogue computer development in Germany

In Germany in 1939, Helmut Hoelzer, a graduate in Electrical Engineering from the Technical University of Darmstadt, was among those drafted to Werner von Braun's long-range rocket project at Peenemunde. He began work as a member of a team assigned the task of developing a radio guidance system for the A-4 (commonly known as the V-2) rocket.[4] Building on an idea he had as a student in Darmstadt, for using capacitors to achieve integration and differentiation, Hoelzer helped develop an electronic guidance computer. Having worked on this special-purpose device, Hoelzer decided to try to design a more versatile electronic analogue computing system that could be used to simulate a missile and its guidance system. By 1941 Hoelzer had completed the construction of a working prototype and was authorised to formalise the design. During the war this computing system was used to simulate the A-4 rocket in two degrees of freedom.[5]

Hoelzer was among the members of von Braun's rocket team who were transported to the USA in 1946 under Operation Paper Clip.[6] One of Hoelzer's electronic analogue computers was also shipped to the USA. At Fort Bliss in the late 1940s Hoelzer's electronic analogue computer was employed in simulation studies of the US Army's Hermes rocket. In late 1949 the Army's missile programme was transferred from Fort Bliss to the Redstone Arsenal in Huntsville, Alabama, and the Guided Missile Development Division was established. This was the US Army's main ballistic missile research and development centre. At Huntsville, in 1950, the construction of a second larger analogue computer was completed. This was based on, and constructed in part from, components of the original system designed by Hoelzer during the war. This computer was used in simulations of the Army's Redstone and Jupiter ballistic missiles and remained in service for ten years.[7]

Wartime developments in Britain and the USA

In Britain and the USA World War II initiated an enormous expansion of research and development funding and activity. The drive to develop radar, gun-fire controllers and training simulators led to accelerated developments in electronic devices, electrical components, circuit techniques and servomechanisms.

An important British centre for research and development work in these areas was the government-funded Telecommunication Research Establishment (TRE) at Malvern. At TRE during the war, a group headed by G. W. A. Dummer was responsible for the design and development of air crew trainers for all of the new airborne radar systems that went into service.[8] In 1941 Dummer and his staff designed an electronic analogue system that simulated a simplified version of the aerodynamics of fighter aircraft. This system was developed as part of an aircraft interceptor simulator that was used to train fighter pilots and crew in the use of night-time radar to intercept bombers. The electronic simulator was based on AC amplifiers and used an electromechanical integrating device known as a Velodyne. This device was developed at TRE by A. M. Uttley and F. C. Williams.[9] At TRE work on the development of electronic and electromechanical analogue computing devices and systems was also undertaken in connection with the development of anti-aircraft gun controllers. One such device was the No. 11 Anti-Aircraft Gun Predictor, developed for use by the British Army.

At the end of the war the transfer of staff from TRE to other government research establishments and to the aeronautics industry led to the dissemination of wartime developments. For example, after the war L. H. Bedford, who had worked on the development of the No. 11 Predictor, left TRE and was appointed Chief Engineer at the Guided Weapons Division of English Electric. At English Electric, in 1949 Bedford and his design team developed a device known as the Kinematic Homing Simulator; this was built largely of components from No. 11 Predictors.[10] As we shall see in chapter 4, work at TRE also influenced the postwar development of general-purpose electronic analogue computers at the Royal Aircraft Establishment, Farnborough.

Developments in the USA: the roles of Bell Telephone Laboratories and Columbia University

From the point of view of postwar electronic analogue computing, the wartime work on DC amplifiers at the Bell Telephone Laboratories was pivotal. This work began in 1940 shortly before the USA entered World War II. At that time D. B. Parkinson and C. A. Lovell were working at

Bell Laboratories on the improvement of an automatic level recorder and had no prior experience of gun director systems.[11] Their involvement in gun-controller design had a rather unconventional and sudden beginning. One night Parkinson dreamt of a fantastic anti-aircraft gun which brought down an aircraft with every shot. He saw vividly that the mechanism directing the gun was the same as the control potentiometer that was being used in the automatic level recorder on which he had been working. As a result of the dream he realised the direct analogy between the control of the high-speed motion of the recorder pen and that of an anti-aircraft gun. Parkinson related this experience to his supervisor Dr C. A. Lovell. Along with a colleague, B. T. Weber, they began a study of the possibilities. Together they wrote a series of dissertations on what they called "Electrical Mathematics" in which they detailed the electronic and electrical analogue computing elements required to construct a gun director, including devices to add, subtract, multiply, integrate and differentiate.[12]

In June 1940, Parkinson and Lovell presented their preliminary design proposal to the management at Bell, and later that month to the US Army Signal Corps. The Army were already concerned that in wartime conditions there might be a shortage of sufficiently skilled labour to manufacture their mechanical gun-directors. The Army recognised the significance of the new design and realised that electronic control could replace the precisely machined components of the mechanical directors.[13] Funding for the project was soon approved, and the National Defense Research Committee (Section D-2) awarded the contract to develop the electronic gun-director to Western Electric/Bell Telephone Laboratories in November 1940.[14] The director was designated the T-10 and was to control a standard 90 mm anti-aircraft gun. It was later renamed the M-9 following standardisation by the US Army.

The design of accurate servomechanisms for function generation and gun positioning was one of the most difficult parts of the design. For this purpose, DC operational amplifiers were developed that could handle both positive and negative DC voltages, had well-defined and controlled amplification characteristics and suffered minimal zero drift.[15] To meet these design specifications, highly accurate and stable components, such as the capacitors and resistors used in operational amplifier feedback networks, had to be developed.

Nevertheless, the wartime climate led to rapid progress in all aspects of the design, and a prototype system was ready for testing by late 1941. By the beginning of 1943 the first production models were being deployed in the field. The M-9 contributed much to the war effort: for example, in 1944, during the Allied defence of Antwerp of the 4883 V-1s or "Flying Bombs" fired upon, 4672 were shot down.[16]

The T-10 (M-9) system was not the only gun director under development at Bell Telephone Laboratories in the early 1940s. In 1941, the NDRC instigated a programme of research at the Bell Laboratories, to develop an anti-aircraft gun director utilising AC voltages rather than DC voltages.[17] The use of AC voltage signals was intended to overcome problems of voltage drift which built up in DC operational amplifiers. The prototype system which resulted, the T-15, was reported to be better than the M-9. However, it never entered production as manufacture of the M-9 had already begun.[18]

Though a great deal of work relating to electronic analogue devices and systems was carried out at BTL during the war, none of this was oriented towards the development and construction of a general-purpose electronic computer. Only in the last few months of the war did staff at BTL begin to take an interest in utilising the DC electronic analogue technology of the M-9 gun director for computation rather than control applications.[19] This led to the construction of Gypsy–BTL's General-Purpose Analogue Computer, also known as GPAC. The principal architect of Gypsy was Emory Lakatos. Lakatos gained his experience in analogue computing during the war while working on fire-control projects with the Physical Research Department at BTL and as a consultant to the NDRC. In 1946, during the planning for the general-purpose computer, Lakatos envisaged a machine that would "aid the thinking of the Laboratories' engineers". He felt that potential users at BTL were more interested in gaining an understanding of a given problem, or finding an optimum solution, than solving it for only one set of parameters. This led Lakatos to define the chief design criteria of the new machine to be "its great convenience in use" and an ability to produce new solutions every three to five minutes.

If the general-purpose analogue computers were to be a valuable research and design tool, then for Lakatos, flexibility and a user interface which enabled the engineer to vary parameters were central considerations. The desire to allow easy interconnection of computing elements meant that it would not be possible to accomplish the economy of apparatus achieved in the M-9 gun-director computer. Flexibility required that the input and output terminals of each DC operational amplifier be brought out to a switchboard where interconnections were to be made using patch cords similar to those used on manual telephone exchanges.

By 1949 the construction of the first of two Gypsy analogue computers was completed. Gypsy was able to solve twelfth-order differential equations to an accuracy of between 0.1% and 1%, with the results appearing in graphical form on an electrically driven plotting board. In addition to the solution of linear and non-linear differential equations encountered in the design of telephone relay systems, Gypsy was used to solve simultaneous linear equations with twelve variables.[20] In the postwar boom in military-funded R&D at BTL, Gypsy was employed in many military-related applications. It was, for example, used during the US Navy Intercept Project, established at BTL in June 1949 and in the 1950s by the Military Electronics Department at BTL.[21]

Gypsy, however, was never manufactured on a commercial basis. The most significant contribution made by BTL to the postwar electronic analogue computer industry was the pioneering work on DC amplifier design undertaken there during the war. This work established much of the technological basis for the subsequent development of commercial electronic analogue computing. Yet as we have seen, its primary goal was not the development of a general-purpose electronic analogue computer; this was a subsequent development. Lovell and Parkinson enrolled DC amplifiers and redefined them as computing devices that solved fixed, predefined mathematical formulations associated with the automatic control of an anti-aircraft gun. The first concrete steps to redefine and codify DC operational amplifiers into a generalised analogue computing system, and to enrol them as aids to computation rather than a device for use in single-purpose simulators and control applications, were taken within a wartime project at Columbia University in the USA.

The Columbia University project was established in 1943 and was one of many sponsored and organised during the war by Section 7.2

(Airborne Fire Control) of the National Defense Research Council. The project's principal concern was the investigation of designs for new airborne bomb navigation devices and gun-fire controllers. For practical reasons many of the designs had to be evaluated in the laboratory, and thus methods of doing so had to be developed. One of the wartime researchers at Columbia, John R. Ragazzini, later remarked: "The impracticality of carrying out controlled experiments in the field was self-evident, particularly since many of these systems existed only on paper and had to be tested in the laboratory."[22] Unable to use in-the-field trial-and-error methods, Ragazzini and his colleagues, Robert H. Randall and Frederick A. Russell, enrolled feedback amplifiers to reconstruct dynamic systems as electronic models and to substitute bench-top simulations for experimental field-trials.

Experiments using DC operational amplifiers for this purpose began after Professor J. B. Russell saw the circuit diagrams of BTL's M-9 anti-aircraft gun director and brought the DC amplifier techniques used there to the attention of the researchers on the Columbia University project.[23] During the course of the research, several small electronic analogue computing devices were built and used to study the dynamics of the physical systems under investigation. For reasons of national security, details of the work at Columbia University were not published until 1947. In this publication the authors described the characteristics of DC operational amplifiers and established some of the techniques for their use as aids to computation. The paper was entitled "Analysis of Problems in Dynamics by Electronic Circuits" and it is here that the term "operational amplifier" is used for the first time.[24]

The work at Columbia University and BTL associated with DC operational amplifiers played different, yet to an extent overlapping, roles in the history of electronic analogue computing. BTL did pioneering work on the development and use of DC feedback amplifiers as operational amplifiers in control applications. The work at Columbia University on the other hand helped establish the idea that DC operational amplifiers could be viewed as a computational tool-kit, that they could be used to solve equations and to model and simulate dynamic systems. It also indicated the type of applications to which they were suited.

GENERAL-PURPOSE ELECTRONIC ANALOGUE COMPUTERS: FORM AND FUNCTION

The construction and commercialisation of the first general-purpose electronic analogue computer systems took place in the USA in the late 1940s and in Britain in the early 1950s. These systems emerged from several different, largely independent, projects associated with the development of aircraft and missiles. These systems established much of the form and function of the postwar electronic analogue computer. Most of the commercial systems contained between 15 and 25 computing amplifiers, but were modular so that several computers could be combined to solve larger problems. Much of the appeal of electronic analogue computers was that they could compute with great speed: computing elements operated in parallel and problems were solved in real, or a fraction of real, time.

Electronic analogue computers can be divided into two distinct traditions: the "slow" or "single shot" and the "repetitive operation" (rep-op) systems. Briefly stated, in a single-shot computer the problem is set up and run, and a solution plotted on paper; parameters may then be varied and the problem re-run. On a repetitive-operation computer the problem was automatically re-run and the solution displayed on an oscilloscope. On the early rep-op computers the automatic repetition rate could be set at between 10 and 60 times per second. Parameter variations could be made during the run and the effects observed almost immediately. This made repetitive-operation computers particularly well suited to engineering design problems involving parameter-variation methods and optimisation. However, the solutions produced by high-speed repetitive operation were less precise and accurate than those from the single-shot operation. In the early 1950s the majority of commercial electronic analogue computers were single-shot systems. However, by the late 1950s considerable improvements in electronic components and circuit designs led to improvements in performance of repetitive-operation computing. Subsequently, manufacturers of single-shot computers converted their equipment to operate in either single-shot or repetitive-operation.[25]

Postwar general-purpose electronic analogue computers consisted essentially of three types of equipment: a) equipment to enter data, generate forcing functions and set up (or program) the computer, b) computing equipment to perform the computation, and c) equipment to record or display output data. In single-shot and repetitive-operation computers, DC operational amplifiers were the principal computing components.[26]

Alternating or Direct Current amplifiers
The parallel development of both AC and DC amplifiers continued well after the end of World War II, and it was not self-evident which of the two amplifier technologies and techniques would dominate in electronic analogue computing applications.

The main disadvantage of the AC operational amplifier as an analogue computing element was that integration could not be performed directly but had to be done by electro-mechanical devices. DC operational amplifiers, on the other hand, perform integration, differentiation and addition directly. It is a fairly straightforward process to configure these devices to perform such operations since it requires only the selection and arrangement of a few external components, namely resistors and capacitors. The main advantage of the AC operational amplifier was that it did not suffer from the voltage drift problems encountered with the early DC operational amplifiers. Voltage drift was particularly troublesome for problems with relatively long solution times, largely because erroneous voltage levels could build up on the input of the amplifier, leading to significant errors in the overall solution. These problems were overcome with the development of a range of techniques to stabilise DC operational amplifiers. The most widely adopted of these was the chopper stabilisation method developed in the late 1940s by E. A. Goldberg while working at RCA on the US Navy's Typhoon analogue computer project. As we shall see in chapter 4, projects funded by the US Navy played a formative role in the postwar development of electronic analogue computing. As a result of the development of chopper stabilisation methods, the contest between AC and DC was largely resolved in favour of DC devices and techniques.[27]

Figure 3.1 illustrates the symbols used to describe DC operational amplifiers and shows how they can be configured to perform either inversion/scaling, addition, integration or differentiation.

The diagram demonstrates how a number of operational amplifiers can be combined to solve a second-order differential equation describing the motion of a weight suspended on a spring without damping. To set up an analogue computer to solve this problem

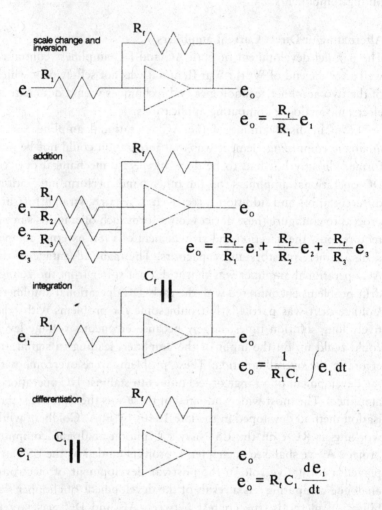

Operational amplifier configurations   Mathematical operation

scale change and inversion

$$e_o = \frac{R_f}{R_1} e_1$$

addition

$$e_o = \frac{R_f}{R_1} e_1 + \frac{R_f}{R_2} e_2 + \frac{R_f}{R_3} e_3$$

integration

$$e_o = \frac{1}{R_1 C_f} \int e_1 \, dt$$

differentiation

$$e_o = R_f C_1 \frac{de_1}{dt}$$

FIGURE 3.1 EXAMPLES OF THE USE OF OPERATIONAL AMPLIFIERS AS ANALOGUE COMPUTING ELEMENTS

would require only a few connecting wires and components. However, problems were usually considerably more complex, involving dozens or even hundreds of operational amplifiers, as well as other computing components, and a great many interconnections had to be made. To organise, simplify and make the process of setting up the analogue computer more efficient, a system of patch panels

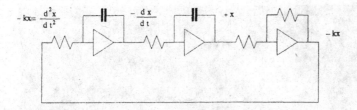

Equation of motion of a weight suspended on a spring

$$a\frac{d^2x}{dt^2} + b\frac{dx}{dt} + kx = 0 \qquad \text{where } a = 1, b = 0$$

$$\frac{d^2x}{dt^2} = -kx$$

FIGURE 3.2 EXAMPLE OF THE USE OF ANALOGUE COMPUTER ELEMENTS TO SOLVE A SECOND ORDER DIFFERENTIAL EQUATION

and/or removable patch boards was developed. The patch panel contained a matrix of connection sockets which corresponded to the matrix of input, output and control signal terminations brought out to a panel on the computer itself. These panels were usually located on the front of the computer for ease of access. The patch panel was used to set up the specific pattern of interconnections of computing amplifiers in the analogue computer for a given problem. Interconnections between components were made via cables inserted into connectors on the face of the patch panel or patch board which was then inserted into the patch panel. Removable patch boards had the advantage that they could be prepared away from the computer and after use removed and stored, see Figure 3.3. By the mid-1950s most of the leading commercial electronic analogue computers were equipped with centralised patch panels with removable patch boards.

Among the other major components usually located on the front panel of the analogue computer were the rows of potentiometers. These were used during set-up and the computation to set and modify the values of coefficients in the equations being solved. Besides the electronic computing amplifiers, general-purpose analogue computers also contained various other computing elements. Generally these included units to generate linear and non-linear functions and to perform multiplication and division. An enormous

FIGURE 3.3   REMOVABLE PATCH BOARD [28]

variety of devices, both electronic and electro-mechanical, were
developed to perform these functions. Often these computing devices
were housed in separate cabinets. Thus, a large analogue computer
system generally consisted of a number of interconnected cabinets
configured to provide the range of computing elements that was
required. There follows a brief description of two of the principal
auxiliary computing devices: multipliers and function generators.

Multiplication

Multiplication could be performed in a number of ways. DC operational amplifiers could only multiply by a constant; when the
product of two variables was required, either servomechanical or
electronic multipliers were used. The servomechanical multiplier consisted of a set of ganged potentiometers driven by a servo-motor to a

position determined by the input voltage to the servo-system. Electronic multipliers, of which there were a great many different types, such as quarter-square diode and time-division multipliers, had a greater frequency response than the servomechanical devices. However, the early rudimentary electronic multipliers were generally less accurate. The development of more accurate electronic multipliers was a constant feature of the overall development of postwar electronic analogue computing.

Function generation

Function generators took many forms. Cathode ray tube-based devices were among the earliest and most widely used function generators. Briefly stated, these translated complicated functions into electrical voltages that the analogue computer could operate on, in the following manner. An opaque mask was cut to correspond to the curve of a function $y = f(x)$. This was then placed on the front face of the cathode ray tube (CRT), and a light-sensitive device known as a photo-electric cell was placed in front of the CRT. As the electron beam from the CRT scanned across the face of the tube, the photo-electric cell detected changes in light levels caused by the edge of the

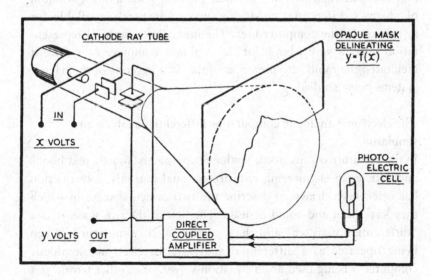

FIGURE 3.4 CRT BASED FUNCTION GENERATOR [29]

mask and generated a signal. This signal was fed back to the CRT in such a manner as to keep the electron beam just visible as it followed the outline of the mask. As the beam followed the function, it produced a voltage waveform equivalent to y = f(x) which was used as an input to the analogue computer, see Figure 3.4. Function generators based on the characteristics of diodes were also developed. If a function could be represented by a sequence of straight-line segments, then an equal number of diode circuits could be used to construct an approximation to that function, based on their near-straight-line voltage transfer characteristics. Operational amplifiers and tapered potentiometers and servomechanisms were among the other methods of function generation.[30]

The principal devices used to measure and record output data were voltmeters, oscilloscopes and plotters. Voltmeters were used to take readings at various points between the interconnected computing elements. These voltages corresponded to steady-state values of parameters associated with the solution of the differential equations. Oscilloscopes and plotters were used to display or record the dynamic components of the analogue computer solution. These devices presented data from the analogue computer in a graphical form. The oscillation or curve that represented the transient behaviour of a particular part of a dynamic physical system, or the solution of abstract differential equations, was immediately available for inspection by the computer user. The immediate graphical representation of data was a boon for the analogue computer user since it facilitated the rapid interpretation of the data in terms of the overall systems being studied.

The electronic analogue computer as differential analyzer or simulator

In the literature on electronic analogue computers, mostly text books and articles in engineering and professional journals, a distinction has often been drawn to describe the two principal ways in which they have been and could be used. The first of the two ways is as a "differential analyzer" and the second is as a "simulator". When being operated as a differential analyzer, the electronic analogue computer is being used as an equation solver, or in other words as a mathematical tool. When operated as a simulator, it is being used to construct an analogous electronic model, generally of dynamic physical systems, and to simulate its behaviour under a range of operating conditions.[31]

These different uses have also been associated with, and used to reflect, differences in engineering and scientific practice:

> ... [consider] the different ways in which two men—a mathematician and an engineer—might attack some problem concerning a dynamic system. The mathematician would examine the system, write down the differential equations of motion, and then build—or have built for him—a differential analyzer to solve the equations. The engineer would examine the system and would then build a model of it, which he would call a simulator. He could then obtain solutions of his problem, perhaps without ever having written down the full set of equations of motion. Thus both men would obtain the required solutions, by somewhat different thought processes, but quite possibly the two calculating machines would be identical.[32]

In practice, electronic analogue computers were used less as differential analyzers and more as model-building systems for experimentation and simulation studies in engineering applications.[33] Historically, the appeal of analogue computing devices was, and still is to an extent, their correspondence with the physical world. Techniques involving the construction of models, either mathematical or physical, of real-world systems have been and still are an invaluable aid to science and engineering. Yet, though all models can be said to be analogous to the system under study, mathematical models need not be implemented or solved using analogue computing devices. Numerical methods and digital computers are extremely versatile and powerful tools which can also be used for mathematical modelling and simulation.

ELECTRONIC ANALOGUE AND DIGITAL COMPUTERS : A COMPARISON

Vis-à-vis digital computers, the electronic analogue computer has historically been seen as having a number of advantages; relatively low cost, high-speed computing and the intrinsic appeal of analogue computing technology and technique. This appeal was based largely on the close correspondence between the dynamic physical system under study and the analogue representation or model, as well as the use of hands-on, interactive computing techniques and the graphical representation of data. In general, their main disadvantage was that they were not high-precision computers; the precision of the solution was limited by the precision and tolerance to which the physical computing components could be manufactured. Initially, systems operated to an accuracy of between 2% and 5%; by the early 1950s this had been improved to 0.1%.

Fundamentally, electronic analogue computers were parallel-processing machines in which a separate computing element was allocated for each arithmetic and mathematical operation. The use of electronic devices and parallel operation permitted high computing speeds. Parallel digital computers have also been developed, but most digital computer systems are based on the von Neumann architecture, and thus operated in a sequential (or serial) manner. Though the components of digital computers operate at extremely high speeds, the combination of serial architecture and numerical computing methods meant that until the early 1960s most digital computers were too slow to be useful in real-time studies of dynamic systems.

However, one disadvantage of analogue parallelism was that as a problem grew in size and complexity, more computing components were required (amplifiers, multipliers, etc.) and the analogue computer also had to grow. With the digital computer it is generally the case that if a problem grows in size and/or complexity, then the time to solve it increases, rather than the size of the computer. The nature of analogue computation and the competition between analogue and digital computing are issues that are discussed at greater length in chapter 8. In the following chapter the central and formative influence of military programs and projects on the development and commercialisation of the postwar electronic analogue computer is explored.

NOTES

1. My use of enrolment follows that of Callon and Latour, but I differ from the former in that I do not see this as a symmetrical process. See, M. Callon, " Mapping the Dynamics of Science and Technology: The Case of the Electric Vehicle", in: Callon, M., J. Law and A. Rip (eds) *Mapping the Dynamics of Science and Technology: Sociology of Science in the Real World*, Macmillan Press, London, 1986; B. Latour, *Science in Action*, Open University Press, Milton Keynes, 1986.

2. P. A. Holst, "George A. Philbrick and Polyphemus—The First Electronic Training Simulator", *Annals of the History of Computing*, Vol. 4, no. 3, April 1982, pp. 143–156.

3. K. L. Wildes and N. A. Lindgren, *A Century of Electrical Engineering and Computer Science at MIT: 1882–1982*, MIT Press, Cambridge, Mass., 1985, p. 188.

4. J. E. Tomayko, "Helmut Hoelzer's Fully Electronic Analogue Computer", *Annals of the History of Computing*, Vol. 7, no. 3, July 1985, pp. 227–240.

5. Ibid.

6. T. Bower, *The Paperclip Conspiracy: The Battle for the Spoils and Secrets of Nazi Germany*, Michael Joseph, London, 1987.

7. H. H. Hosenthien and J. Boehm, "Flight Simulation of Rockets and Spacecraft", in E. Stuhlinger, F. I. Ordway, III, J. C. McCall and G. C. Bucher, (eds.), *From Peenemunde to Outer Space*, NASA George C. Marshal Space Flight Centre, Huntsville, 1962, pp. 437–469.

8. G. W. A. Dummer, "Aids to Training—The Design of Radar Synthetic Training Devices for the RAF," *IEE Journal*, Vol. 96, Pt. III, 1949, pp. 101–116.

9. F. C. Williams and A. M. Uttley, "The Velodyne", *IEE Journal*, Vol. 93, Pt. III A, 1946, pp. 1256–1274; F. C. Williams, "Electro Servo Simulators," *IEE Journal*, Vol. 94, Pt. II A, 1947, pp. 112–129.

10. I. N. Cartmell and R. W. Williams, "Guided Weapons Simulators", *Journal of the Royal Aeronautical Society*, Vol. 72, April 1968, pp. 356–360.

11. W. H. Higgins, B. D. Holbrook and J. W. Emling, "Defence Research at Bell Laboratories: Electrical Computers for Fire Control.", *Annals of the History of Computing*, Vol. 4, no. 3, July 1982, pp. 218–236.

12. C. A. Lovell, D. B. Parkinson and B. T. Weber, US Patent, No. 2,404,387, filed May 1, 1941; issued July 23, 1946; C. A. Lovell, "Continuous Electrical Computation", *Bell Laboratories Record*, Vol. 25, March 1947, pp. 114–118; Anon., "Development of the electrical director," *Bell Laboratories Record*, Vol. 22, May 1944, pp. 225–230.

13. Anon., "Development of the electrical director," p. 221.

14. The Western Electric Co was responsible for overseeing contractual matters for Bell Laboratories at that time.

15. The first explicit use of negative feedback to stabilize the amplification characteristics of thermionic valves was undertaken in connection with the design and improvement of communication networks. During the Second World War another BTL employee, Hendrik W. Bode, established the mathematical methods for designing feedback amplifiers to operate within specific tolerances. H. Nyquist. "Regeneration Theory" *Bell System Technical Journal*, Vol. 11, Jan 1932, pp. 126–147; H. S. Black, "Stabilised Feedback Amplifiers", *Bell System Technical Journal*, Vol. 13, Jan 1934, pp. 1–18; H. W. Bode, *Network Analysis and Feedback Amplifier Design*, Van Nostrand, New York, 1945.

16. C. Eames and R. Eames, *A Computer Perspective*, Harvard Univ. Press, Mass., 1973, p. 128; J. P. Baxter, *Scientists Against Time*, Little, Brown, Boston, Mass., 1946.

17. Following the initial proposal from Bell Telephone Laboratories several other firms were also engaged by the NDRC to investigate the use of electrical and electronic analogue methods across the spectrum of gun-fire control devices. The Radio Corporation of America and the Arma Corporation were amongst the firms that undertook research in this area. For example, at the Arma Corporation an electronic analogue controller was developed for the Mark 47 anti-aircraft gun. See B. Rowlands and W. B. Boyd, *US Navy Bureau of Ordnance in World War II*, Bureau of Ordnance, Department of the Navy, US Government Printing Office, Washington, DC, 1953. Flight trainers were also developed at BTL during the war most notably a trainer for the US Navy's PBM-3 aircraft was completed in 1943. This used a fixed-purpose electronic analogue computing system to solve the flight equations. In total, 32 trainers for 7 aircraft were developed at BTL during the war. See, Anon., "Operational Flight Trainer Uses 200 Tubes," *Electronics*, Feb 1945, pp. 214–216; L. de Florez, "Synthetic Aircraft", *Aeronautical Engineering Review*, April 1949, pp. 26–29.

18. W. H. C. Higgins, B. D. Holbrook and J. W. Emling, "Defence Research at Bell Laboratories: Electrical Computers for Fire Control", p. 229.

19. A. A. Currie, "The General Purpose Analog Computer", *Bell Laboratories Record*, Vol. 29, no. 3, March, 1951, pp. 101–107. See Bell Telephone Laboratories, *A History of Engineering and Science in the Bell System: National Service in War and Peace (1925–1975)*, Bell Telephone Laboratories, 1978; M. G. Stevenson, "Bell Labs: A Pioneer In Computing Technology: Part 1. Early Bell Labs Computers", *Bell Laboratories Record*, December 1973, pp. 345–350, p. 349. In a series of memoranda in late May and early June 1945, staff at Bell Labs began a discussion on the possible use of analogue computing elements developed during the war for electrical gun directors as the nucleus for an electrical differential analyzer. By late June of the following year, engineers at BTL and in particular Emory Lakatos realised that a "machine of far greater capabilities than was originally anticipated or than can be realized in a standard differential analyzer" was possible. To emphasise the greater breadth of applications which the new machine could tackle, Lakatos recommended that it no longer be referred to as a differential analyzer, but instead called it a "general analog computer." E. Lakatos, "Interim Report on the Electrical Analog Computer", *Internal typescript Memorandum MM-46-10-60*, June 28, 1946, pp. 1–26. AT&T Archives, Case 20878; E. Lakatos, "Problem solving with the analog computer", *Bell Laboratories Record*, Vol. 29, no. 3, 1951, pp. 109–114.

20. E. Lakatos, "Problem solving with the analog computer".

21. It was used in attack phase simulations of guided missiles (i.e. when the missile is operating on data from its own radar). Anon, *Navy Intercept Project Quarterly Engineering Report No. 18*, 1 Oct 1953–1 Jan 1954. AT&T Archives, Case 26656-100. And in studies on the development of a fully automated carrier-based aircraft landing system. Anon., *Automatic Carrier Landing Study: Final Report*, April 1954. AT&T Archives, Case 26656-89.

22. J. R. Ragazzini, R. H. Randall and F. R. Russell, "Analysis of Problems in Dynamic by Electronic Circuits", *Simulation*, Sept 1964, pp. 54–65, p. 54. This is a reprint of the original article with foreword by John R. Ragazzini. Original published in *Proc., IRE*, Vol. 35, May 1947, pp. 444–452.

23. The circuits were taken from the instruction booklet prepared by Bell Laboratories for the Western Electric M-9 antiaircraft gun director. J. R. Ragazzini, R. H. Randall and F. R. Russell, "Analysis of Problems in Dynamic by Electronic Circuits", *Proc., IRE*, Vol. 35, May 1947, pp. 444–452.

24. Ibid.

25. H. M. Paynter (ed.), *A Palimpsest on the Electronic Art*, George A. Philbrick Researches Inc, Boston, Mass., 1955; R. Tomovic and W. J. Karplus, *High-speed Analog Computers*, John Wiley and Sons, New York, 1962; C. L. Johnson, *Analog Computer Techniques*, 2ed., McGraw-Hill, New York, 1963.

26. G. W. Smith and R. C. Wood, *Principles of Analog Computation*, McGraw-Hill, New York, 1959.

27. The use of AC operational amplifiers as analogue computing elements did not, however, come to an abrupt halt. In the 1950s they continued to be used in specialized application where the DC operational amplifiers was considered an unproven technology. This was most evident in the field of special-purpose electronic analogue computers for in-flight simulators and air crew trainers.

28. *EAI 231R-V Information Manual*, Electronic Associates Inc., Long Branch, NJ, 1964, p. 6.

29. D. J. Mynall, "Electrical Analogue Computing—Part 3—Functional Transformation", *Electronic Engineering*, Vol. 19, Aug 1947, pp. 259–262, p. 262.

30. R. E. Hare and W. E. Willison, "Analogue Computers", Part 2, *Instrument Practice*, Dec 1958, pp. 1295–1303.

31. W. Soroka, *Analog Methods in Computation and Simulation*, McGraw-Hill, New York, 1954.

32. C. A. A. Wass, *Introduction to Electronic Analogue Computers*, Pergamon Press, London, 1955, p. 5.
33. When the starting point was a set of differential equations. Before these could be set up on the computer they first had to be "scaled". Scaling was necessary to ensure that the maximum values of coefficients would not exceed the dynamic range of the computer voltages. However, in the absence of a rigorous mathematical definition it was still possible to proceed. The development of block diagram oriented programming systems, enabled computing elements, defined as specific operations, to be combined to model a dynamic system without necessarily referring to, or defining, the mathematical equations.

CHAPTER 4

Electronic Analogue Computer Development, 1945–1955: Military Programmes, Aeronautics and Electronics

INTRODUCTION

In a review of analogue computer developments published in 1958, the authors, R. E. Hare and W. E. Willison, noted:

> The design of advanced equipments for military use inevitably made increased demands on aids to mathematical computation. This was particularly true in the analysis of complex closed-loop control systems: the need for analogue computers was established, and it is probably true to say that without them progress in the aviation and guided missile fields would have been considerably retarded.[1]

In the USA and Britain, postwar military programmes and the design of equipment for military use established a variety of economic and technical imperatives that led to the design and construction of a new, alternative, computing technology—the general-purpose electronic analogue computer. The postwar construction of the electronic analogue computer involved the redefinition and enrolment of electronic devices that had been developed largely during the war for fixed-purpose automated control and simulation applications. What emerged from this work was a unified computational technology and set of techniques that became a valuable tool for engineering design. Yet this postwar transformation was not led by the prewar centres of analogue computing, but was undertaken by the electronics and aircraft industries, in projects funded by military agencies, and in government research establishments associated with the design and testing of missiles, aircraft and other weapon systems.

The chief economic imperative for developing electronic analogue computers was to reduce the enormous costs arising from the development of aircraft and missile systems. A major component of these costs stemmed from the use of experimental field trials with full-scale prototype systems as a way of obtaining performance data. Missile firings were particularly expensive because the missile was generally destroyed in the process. Clearly, the fewer field trials required to validate the effectiveness of a weapon system, the greater the savings. Yet in the 1940s and 1950s a great many aspects of the control and dynamics of missiles and aircraft were not well understood. Moreover, the nature and complexity of these technical systems meant that they were not amenable to design by existing analytical or empirical methods. Without the aid of high-speed computers, analytical techniques were too time-consuming. Existing empirical trial-and-error and parameter-variation methods were based on the use of physical scale models and field trials with full-scale systems. Yet experimental data from scale models was only of limited use, and problems arose in scaling-up. Moreover, design methods requiring a large number of experimental test flights needed to be avoided, not only because of the costs involved, but also because of safety and national security considerations. Funded by governments and military agencies and driven by the computational demands arising out of military programmes, high-speed electronic analogue computers were constructed to bring the missile and aircraft into the laboratory, where it could be designed, modelled and simulated—or flown—hundreds of times, safely, in secrecy and relatively cheaply.

Indeed, in the USA and Britain the role of government and military-funded projects in the development of electronic analogue computers was pivotal. In this chapter we shall see that in the USA the Navy and Air Force established and directly funded projects in private industry to develop general-purpose electronic analogue computers. In Britain, as we shall see in chapter 6, the bulk of the formative postwar research and development in this area was undertaken at the Royal Aircraft Establishment, Farnborough—a government-funded centre for guided missile and aircraft development.

In the USA and Britain, military projects and related weapons programmes not only laid the technical and institutional foundations of postwar electronic analogue computing, but also sustained subsequent development work in analogue computing in two major ways. Firstly, the armed forces purchased commercial systems for their own research laboratories and test facilities. Secondly, firms working on the development of weapon systems for the military either developed or purchased electronic analogue computers for use in the design and testing of these systems. Moreover, in the USA the Office of Naval Research (ONR) organised symposia and played a role in disseminating information on new electronic analogue computer applications and on developments in the theoretical and practical techniques employed in their use.

Before we look at the analogue computer situation in more detail it is important to point out that the military, especially in the USA, also had an enormous influence on the development of digital computing. Post-World-War-Two national security concerns, arising out of the "Atomic Age" and Cold War politics pushed on the development of control and communication systems, and air defence and early warning systems on the ground, at sea, and in the air. In the USA, the Navy, the Air Force and the Army all sponsored independent groundbreaking research and development work. The military stimulated development through the detailed specification and purchasing of novel digital computing equipment. It has been estimated, that by 1950 military and federal funding for digital computer R&D amounted to between $15 million and $20 million.[2] As Flamm points out in his study of digital computer development, even a partial list (see Table 4.1) shows the degree of military involvement in the development of large and smaller scale digital computer projects during the 1940s and 1950s.

In this chapter we see that, from 1945 to 1955, the military also played a fundamental role in the origins of electronic analogue computing in the USA. British developments during this period are dealt with in chapter 6.

MILITARY PROGRAMMES AND ELECTRONIC ANALOGUE COMPUTING IN THE USA

In the USA after World War II, the Army, Navy and Air Force initiated largely separate missile and aircraft development programmes.[3]

TABLE 4.1 MILITARY SUPPORT FOR EARLY DIGITAL COMPUTERS IN THE USA

First Generation Computers	Cost $ 1000	Source of Funding	Initial Operation
ENIAC	750	Army	1945
Harvard Mark II (partly electromechanical)	840	Navy	1947
Eckert-Mauchly BINAC	278	Air Force (Northrop)	1949
Harvard Mark III (partly electromechanical)	1,160	Navy	1949
NBS Interim computer (SEAC)	188	Air Force	1950
ERA 1101 (Atlas I)	500	Navy/NSA	1950
Eckert-Mauchly UNIVAC	400–500	Army via Census; Air Force	1951
MIT Whirlwind	4,000–5,000	Navy; Air Force	1951
Princeton IAS computer	650	Army; Navy; RCA; AEC	1951
Univ. of Cal CALDIC	95	Navy	1951
Harvard Mark IV	–	Air Force	1951
EDVAC	467	Army	1952
Raytheon Hurricane (RAYDAC)	460	Navy	1952
ORDVAC	600	Army	1952
NBS/UCLA Zephyr computers (SWAC)	400	Navy; Air Force	1952
ERA Logistics computer	350–650	Navy	1953
ERA 1102 (3 built)	1,400	Air Force	1953
ERA 1103 (Atlas II 20 built)	895	Navy/NSA	1953
IBM Naval Ordnance Research Computer (NORD)	2,500	Navy	1955

Source: K. Flamm, *Creating the Computer: Government, Industry, and High Technology*, The Brookings Institution, Washington DC, 1988, pp. 76–77; RCA: Radio Corporation of America, AEC: Atomic Energy Commission; NSA: National Science Foundation

The rather disparate approach to missile and aircraft programmes as well as differences in the postwar organisation of military research, resulted in different approaches to the development of electronic analogue computers.

In the late 1940s the US Navy directly sponsored four projects involving the development of electronic analogue computers. Two of these were funded through the Office of Naval Research and carried out by electronics firms: Project Cyclone at the Reeves Instrument Corporation, and Project Typhoon at the Radio Corporation of America. Project Cyclone led to Reeves becoming the first American company to manufacture general-purpose electronic analogue computer systems on a commercial basis. Projects Cyclone and Typhoon were instrumental in establishing the technological basis for the postwar general-purpose electronic analogue computer industry in the USA. The US Navy Bureau of Ordnance also funded two projects: one at Dynamic Analysis and Control Laboratory at MIT, and one undertaken jointly by the Curtiss-Wright Corporation and the Applied Physics Laboratory at Johns Hopkins University.

Initially, the US Air Force did not establish large projects of a scope and scale as those sponsored by the US Navy. It did, however, fund the development of a general-purpose electronic analogue computer at the Goodyear Aircraft Co. The USAF also sponsored analogue computer development indirectly as a consequence of placing contracts with aircraft companies for missile systems. One of these was the Boeing Aircraft Co., of Seattle, Washington. Boeing initiated its own project to develop a relatively small analogue computer system. Though they were originally intended for in-house use, both Boeing and Goodyear later manufactured their electronic analogue computer systems on a commercial basis.

In the formative postwar years the US Army contributed little to the development of electronic analogue computing. Initially, the US Army utilised a small special-purpose electronic analogue computing system built in Germany during the war. After the war this equipment had been transported along with the German rocket scientists to the USA, under Operation Paper Clip. Further development of this computer did take place. However, the US Army's chief influence on the electronic analogue computer industry was as a purchaser of computer systems and as a commissioner of new weapon systems from private firms.

In the USA the armed forces were not the only source of Federal funds for analogue computer development. In the late 1940s the National Advisory Committee on Aeronautics (NACA) commissioned the construction of a small one-off electronic analogue computer system, and in the 1950s equipped its Ames and Langley Research Laboratories with large-scale general-purpose analogue and later hybrid computer facilities. NACA's successor, the National Aeronautics and Space Administration (NASA), played an important role as a major purchaser and user of analogue and hybrid computer equipment.

ELECTRONIC ANALOGUE COMPUTERS AND THE OFFICE OF NAVAL RESEARCH

In the USA, Navy-sponsored R&D formed the dominant institutional and technical context within which electronic analogue computation was directed and redefined. The two most significant projects— Cyclone and Typhoon—were organised and funded by the Special Device Centre of the Office of Naval Research, on behalf of the Navy Bureau of Aeronautics.[4] Founded in 1946, the ONR was created to enable the Navy to maintain its contacts with academic and industrial R&D centres built up during the war, and thus to continue to pursue, in its own laboratories and elsewhere under contract, R&D appropriate to naval requirements and objectives. Until the National Science Foundation was established in 1950, the ONR funded more research and development than any other single military or government agency.[5]

Project Cyclone and REAC, the Reeves electronic analogue computer

In 1946 the ONR established Project Cyclone at the Reeves Instrument Corporation of New York. By 1950 Project Cyclone was the largest electronic analogue computer laboratory and development programme in the USA. For the next twenty years Project Cyclone remained in operation, regularly re-equipped with the latest electronic analogue computer systems developed by Reeves.[6]

Cyclone marked a departure from previous developments in electronic analogue computing in several ways. It was designed from the outset to be a general-purpose computing system and thus differed from earlier single-purpose control devices and the simulators built by Philbrick and Hoelzer. It also differed from previous aircraft and radar simulators in that these systems were developed to train air crews and operators. Cyclone was developed primarily as an aid to computation and a design tool, and employed simulation to that end.

Indeed, the overarching aim of the project was to establish a laboratory where R&D on guided missiles could be conducted and "classified problems in other fields [could] be studied and analyzed". The project also had a number of other goals, including the "investigation of new applications for Cyclone equipment", and to undertake and encourage "theoretical research to ascertain and improve the reliability and accuracy of computer solutions." Moreover, the project was to provide a forum in which the "experiences of individual computer groups" could be combined "to make for more intelligent and efficient operation, generally."[7]

The project was located at the Reeves Instrument Corporation's premises in New York. There development work began with the design and construction of a small prototype system. This was built using DC operational amplifiers based on those developed at the BTL in the early 1940s for the M-9 anti-aircraft gun-director computer.[8] It led to REAC (Reeves Electronic Analogue Computer) which became the centrepiece of project Cyclone, and the first general-purpose electronic analogue computing system sold commercially in the USA.[9]

Project Cyclone grew rapidly during the late 1940s and early 1950s, and by 1955 the project had expanded to three separate computing laboratories. The largest of these housed 13 individual REACs and more than 17 auxiliary cabinets of computing equipment, containing a total of 420 computing amplifiers.[10] In keeping with project goals to improve and to verify the accuracy of results, problems were routinely re-run on digital computers. In a typical application—the simulation of a guided missile in three dimensions—the average run time for a single solution on the electronic analogue computer was approximately one minute. The check solution for this problem by numerical methods on an IBM CPC (Card Programmed Calculator) took 75 hours to run; on an Elecom 100 it took from 60 to 130 hours. Moreover, the comparison generally showed an excellent correlation between analogue and digital solutions.[11]

From the late 1940s the computers at Project Cyclone were engaged in the solution of a wide variety of problems, mostly but not exclusively related to the development of guided missile and aircraft systems. Along with problems in navigation and ballistics, Project Cyclone also dealt with problems in engine control and electrical circuit analysis.[12] In addition to military R&D, computing time was allocated to the solution of problems submitted by private firms and universities. On work submitted by external organisations, it was calculated in 1948 that using REAC cut the cost of computation by 95% compared with manual methods.[13]

Though much of the computation and simulation carried out at Project Cyclone was classified, the project played an important role in disseminating information on electronic analogue computers and their application to engineering problems. This was achieved through symposia, organised jointly by Reeves and the Special Device Centre of the ONR. The first Cyclone symposium was held in New York City in March 1951. One hundred and one visitors from forty one organisations attended (see Table 4.2), and over a two-day period

TABLE 4.2 LIST OF PARTICIPATING ORGANIZATIONS, PROJECT CYCLONE SYMPOSIUM 1951

Air Materiel Command, Wright Paterson USAF Base	McDonnell Aircraft Corporation
Applied Physics Laboratory, Johns Hopkins University	Minneapolis-Honeywell Regulator Corporation
Argonne National Laboratory	National Advisory Committee for Aeronautics:
Bell Aircraft Corporation	Ames Aeronautical Laboratory
Bureau of Aeronautics	Langley AeronauticalLaboratory
Canadian Armament Research Establishment	National Bureau of Standards
Chance-Vought Aircraft Corporation	Naval Air Development Centre
Columbia University	Naval Air Experiment Station
Consolidated Vultee Aircraft Corporation,	Naval Air Missile Test Centre
Fort Worth and San Diego	Naval Research Laboratory
Cornell Aeronautical Laboratory	North American Aviation, Inc.
Curtiss-Wright Corporation	Office of Naval Research
Douglas Aircraft Company	Radio Corporation of America
Fairchild Engine & Airplane Corporation	RAND Corporation
General Electric Company	Republic Aircraft Corporation
Glenn L. Martin Company	Sperry Gyroscope Company
Grumman Aircraft Engineering Corporation	The David W. Taylor Model Basin
Hughes Aircraft Company	University of Michigan: Willow Run Research Centre
Jet Propulsion Laboratory, California Institute of Technology	University of Minnesota
Massachusetts Institute of Technology	University of Pennsylvania, Moore School of Engineering
	US Navy Special Devices Centre
	Yale University

twenty papers were presented. Almost all of the organisations represented did in fact own and were actively using REAC computer systems.[14]

Reeves began manufacturing commercial REAC systems in 1948. In 1949 a REAC computer cost $14,320. A standard combination of computing and peripheral units was two REAC computing units, a servomechanism unit, a six-channel recorder and an input-output table this cost approximately $37,000.[15] The demand for REAC computers grew rapidly, and by 1951 more than sixty REAC systems were in use at almost forty locations, including government and industrial laboratories and universities. Among the earliest users of REAC systems were: the Naval Air Missile Test Centre, Point Magu, the Naval Research Laboratory, Washington DC, the RAND Corporation, North American Aviation Inc., the Applied Physics Laboratory at Johns Hopkins University and the University of Minnesota.[16]

Project Cyclone demonstrated that the electronic analogue computer was a valuable tool in the analysis and design of dynamic systems. It helped to disseminate information on electronic analogue computing techniques and aided in the diffusion of the technology itself. The REAC also established that general-purpose electronic analogue computers were a viable and manufacturable commercial product. Reeves continued to develop and manufacture their REAC computer range, and for more than twenty years was one of the leading manufacturers of electronic analogue computer systems in the USA.

Project Typhoon, the Radio Corporation of America and the Naval Air Development Centre, Johnsville

Project Typhoon, like Cyclone, was designed and built for the Special Device Centre of the ONR by a private contractor, in this case the Radio Corporation of America (RCA). Work on Project Typhoon began in 1947 at RCA's research laboratories in Princeton, NJ. For the next three years engineers from RCA's Electronic Computer Section, headed by Arthur W. Vance, worked on the development of electronic analogue computing devices that would make Typhoon the most reliable, accurate and precise systems of its kind yet built. The primary motivation and project goals behind Typhoon were similar to those set down for Cyclone, and while work on the improvement of computing circuits continued, the knowledge gained was disseminated through Typhoon symposia.[17]

RCA, unlike Reeves, did not become a manufacturer of general-purpose electronic analogue computing systems. Nevertheless, the influence of Project Typhoon on the future of the analogue computers was, though less direct, significant. Indeed, in the mid-1950s Project Typhoon was justifiably described as a "veritable fountain head of improved computing elements".[18] One of the most important developments to come out of Project Typhoon was the invention of the chopper-stabilised, or drift-free, DC operational amplifier by E. A. Goldberg.[19] The development of drift-free DC operational amplifiers resolved the competition between DC and AC electronic analogue computing technologies, with DC becoming the industrial standard for general-purpose computer systems.[20]

The first public demonstration of the Typhoon computer was given on 21 November 1950 and was held at RCA's research laboratories in Princeton. RCA's press release began:

> New RCA Electronic Computer Aids U.S. Air Defence; Expected to Save Millions in [the] Design of Guided Missiles. "Project Typhoon" Solves Complex Air Defence Problems for Entire Cities—Works Missile Equations in 60 Seconds That 2 Mathematicians Would Do in 6 Months—Auxiliary Devices Provide 3-Dimensional View of "Dog Fight" Between Missile and Target.[21]

During the demonstration, Typhoon was shown solving a typical air defence problem, in which a high-speed bomber was attacked by a radar-controlled supersonic missile. Commenting on the demonstration, Dr C. B. Jolliffe, the Executive Vice President in charge of the Division at which Typhoon was built, said:

> Complex simulated problems of a complete guided missile system, which other computers are too small or too inaccurate to handle effectively, can be solved by Typhoon. This will enable the design of equipment with a minimum of experiments that would require [the use of] expensive apparatus, such as missiles, aircraft and ships.[22]

The statistics of the Typhoon computer were rather impressive. It contained approximately 4,000 electron tubes and it had an operational precision of one part in 25,000. Setting up (or programming) Typhoon involved the use of one hundred dials and 6,000 plug-in switchboard connectors. Its nine-foot-high cabinets lined a room fifty feet wide by eighty feet long, with the rest of the floor space being taken up by input devices, plotting tables and two novel 3-dimensional output devices.[23] Typhoon was, however, not built entirely of analogue computing elements. To achieve high-speed yet high-precision multiplication, the engineers at RCA developed a new type of electronic multiplier that combined analogue and digital technologies.[24] Thus Typhoon was, though to a limited extent, a hybrid computer system.

In 1952 the final version of Typhoon was moved from Princeton to its permanent home at the US Naval Air Development Centre at Johnsville, Pennsylvania. Typhoon took over from Cyclone as the single largest, and most accurate, electronic analogue computer in the USA. For the next twelve years Typhoon was involved in many different Naval weapons programmes. One of the first was the Navy Intercept Project (NIP) which Western Electric and the Bell Telephone Laboratories (BTL) undertook on behalf of the Navy Bureau of Aeronautics.[25] The NIP project began in June 1949. Its aim was the development of radar and computer equipment to control ship-based surface-to-air guided missile systems. BTL used both its own General-Purpose Analogue Computer (GPAC or Gypsy) and later the Typhoon facility to perform fundamental mathematical analysis and a wide range of simulation and experimental studies related to missile interception and auto-navigational problems.[26] Typhoon was also used in the design and development of many military aircraft, including the joint Navy, USAF, NACA/NASA and North American Aviation rocket-powered X-15 experimental aircraft.[27]

In 1968, Typhoon, which had cost the US Navy $3 million, was dismantled and replaced by a new $1.5 million suit of five transistorised electronic analogue computers. The new computers had twice the computing capacity of Typhoon yet occupied only one-tenth of the space.[28]

US NAVY BUREAU OF ORDNANCE PROJECTS

The Dynamic Analysis and Control Laboratory's Flight Simulator

The third of the large postwar electronic analogue computer projects funded by the US Navy was undertaken by the Dynamic Analysis

FIGURE 4.1 THE TYPHOON COMPUTER INSTALLATION[29]

and Control Laboratory at MIT. This project differed from Cyclone and Typhoon in that it was neither established nor funded by the ONR, but by the US Navy Bureau of Ordnance. Furthermore, it was undertaken in an academic institution rather than private industry, and it combined DC and AC electronic analogue computing in one system.

The Dynamic Analysis and Control Laboratory (DACL) emerged out of the Servomechanism Laboratory, which Gordon Brown founded in 1940. During the war, the Servomechanism Laboratory grew rapidly on projects for the armed forces, one of the first being the development of a director for the US Army's 37mm anti-aircraft gun. Among the wartime projects undertaken for the US Navy was the development of automatic control systems for the stabilisation of the Pelican and Bat guided missiles.

In October 1945, Albert C. Hall, of the Servomechanisms Laboratory, submitted a proposal to the US Navy for a research and development programme to continue wartime work on guided missile control systems. The previous successful development of control systems for the Pelican and Bat missiles encouraged the Navy to fund the programme. In January 1946 a group of engineers left the Servomechanism Laboratory at MIT to establish the Dynamic Analysis and Control Laboratory, under the directorship of Professor A. C. Hall. The same year, work began on the largest postwar analogue computing project at MIT—the DACL Flight Simulator.[30]

The DACL Flight Simulator, like Cyclone and Typhoon, was built largely as a response to the growing complexity of aeronautical design in the postwar years. Higher flight speeds and the design of auto-pilots and automatic guidance systems for missiles emphasised the inadequacies of traditional design methods. Mathematical techniques that linearised equations, and assumptions that simplified the design process were becoming increasingly untenable. This situation led A. C. Hall to comment "the only practical method of solution of most aircraft dynamic problems is by means of computing machines."[31] However, in 1946, according to Hall "… no existing computing machine was well suited to the solution of many urgent problems".[32] Therefore, Hall and his staff decided to build their own.

The Flight Simulator was part of a guided-weapon development project at MIT, known as Project Meteor. The primary objective for the simulator was to provide a facility to test missile guidance systems and auto-pilots on a one-to-one time scale and, at a more general level, to aid in the study of problems associated with aircraft and guided missile design. The Flight Simulator became operational in 1948. It occupied two floors. On the ground floor, the main feature was a 3-axis, gimbal-mounted flight table. This was used to simulate the flight of a missile or aircraft in order to test elements of the control system which were sensitive to angular motion. The flight table was connected to a dedicated analogue computing facility specially designed to solve the differential equations which characterise the motion of an aircraft in flight. On the second floor was the "Generalized Computer", an electronic analogue computer using both DC and AC computing amplifiers. Though designed to meet specific computational demands arising from missile and aircraft simulation studies, the Generalized Computer, as its name suggests, was a general-purpose electronic analogue computer and could be used to solve a wide range of problems. The design of the AC section of the analogue computer was unique.[33]

By the early 1950s one half of the effort of the DACL was devoted to the simulator: a total of some seventy people organised into a Development Division, an Analysis and Evaluation Section and an Operational Section. From 1948 the Generalized Computer was in almost constant use on a variety of trajectory problems, including fully fledged missile simulations in three dimensions. In the early 1950s the computer was operating fourteen hours per day in two shifts.[34] In 1953, William W. Seifert, Associate Director of the DACL, noted that the simulator had enabled major savings in missile design to be achieved. Seifert estimated that the cost of experimental test flights would have been at least one thousand times greater than their machine-computed studies.[35]

In 1950, when Professor A. C. Hall left MIT to become associate director of the Bendix Research Laboratories, the DACL became an interdepartmental laboratory for the departments of Electrical and Mechanical Engineering, with Professor John A. Hrones as the new director.[36] In February 1952 the simulator was put on a self-supporting basis. Preferential treatment of Navy requests for access to the facility came to an end, and it was instead made available to government agencies and their industrial contractors on an equal basis.[37] However, overarching Navy support for the facility continued until the end of 1953 when the US Air Force took over this responsibility. The simulator continued to operate until 1958 when contracts from government and industry were discontinued and the DACL was closed.[38]

CURTIAC, the CURTIss-Wright Analogue Computer

The US Navy Bureau of Ordnance also funded the development of an electronic analogue computer as part of a guided missile project undertaken jointly by the Applied Physics Laboratory at Johns Hopkins University and the Curtiss-Wright Corporation. The missile development programme originated during World War II. Designated "Bumblebee", the programme called for the development of a surface-to-air missile. The postwar work on electronic analogue computers at APL/Curtiss-Wright led to the CURTIAC (CURTIss-Wright Analogue Computer). The CURTIAC was a twenty DC operational amplifier, general-purpose analogue computer similar in design and operation to the Reeves REAC. Though CURTIAC led to improvements in analogue computer design, it was never commercialised.[39]

INDIRECT SUPPORT FOR ANALOGUE COMPUTER DEVELOPMENT: COLLABORATIVE AND CONTRACT R&D, NAVY LABORATORIES AND TEST CENTRES

The Navy contributed to the development of the analogue computing industry in ways other than through the direct funding of large projects. For example, the EASE (Electronic Analog Simulating Equipment) range of commercial computers manufactured by Beckman Instruments Inc. grew out of collaborative research between the University of California at Los Angeles (UCLA) and the Naval Ordnance Test Station (NOTS) based at Inyokern, Chinalake, CA. The Beckman EASE was based upon a computing system, of the same name, developed at UCLA by Dr Tom Rogers, his Ph.D student Louis G. Walters and engineers from NOTS for rocket research and the simulation and testing of ordnance equipment.[40] The first public demonstration of the new range of EASE computing equipment, organised by the Beckman Instruments Inc., was held in May 1952.[41] By the mid-1950s Beckman had established for itself a position as one of the leading manufacturers of general-purpose analogue computing equipment in the USA.[42]

A secondary, but nevertheless significant influence was the fillip given to the industry through sales of analogue computers to commercial companies and universities working on R&D funded by the Navy. For example, in 1952 Electronic Associates Incorporated, which later became the world's largest manufacturer of high-precision analogue and hybrid computer equipment, received its first order for a general-purpose analogue computing system from Westinghouse. It was subsequently used in design studies of the reactor for Nautilus, the US Navy's first nuclear-powered submarine.

The Navy also bolstered the embryonic computer market by purchasing equipment for its own R&D laboratories and test facilities. The Naval Research Laboratory (NRL),[43] Washington, DC, the Naval Air Experiment Station, Philadelphia, Pa, and the Naval Air Missile Test Centre, Point Magu, Ca, were among the first Naval facilities to receive a REAC system.[44] Naval laboratories also bought analogue computing equipment from other manufacturers: up until the early 1950s the main alternatives to the Reeves REAC were the BEAC (Boeing Electronic Analogue Computer) and the GEDA (Goodyear Electronic Differential Analyzer) and the small, inexpensive modular computing components manufactured by George A. Philbrick Researches Incorporated (GAP/R).[45] The Naval Air Material Centre,

Philadelphia, the Naval Ordnance Plant, Indianapolis, the Naval Medical Research Institute and the Naval Underwater Ordnance Station were all users of analogue computing equipment manufactured by GAP/R.[46]

Thus, for a decade after the end of World War II, Navy-funded R&D shaped analogue computing on many levels, from components to commercial systems and from theory to applications. In this way the Navy played a significant part in establishing the institutional and technological context within which the field had grown. Having funded much of the ground-breaking R&D in this area, the role of the Navy diminished in the mid-to-late-1950s, as the burden for innovating new developments shifted to industry. The analogue computer industry established its own computing centres and funded its own research. This research had to be paid for through the sale of equipment and therefore required an expanding market. In this way the influence of the US Air Force and NASA became increasingly important from the mid-1950s as major purchasers of commercial analogue computing equipment, thus sustaining and promoting new growth and development in the analogue computer industry.

THE US AIR FORCE AND ELECTRONIC ANALOGUE COMPUTER DEVELOPMENT

Between 1945 and 1950 the US Air Force, and its predecessor the Army Air Force, played a relatively limited role in electronic analogue computer development. The Army Air Force's role in electronic analogue computer development was fairly indirect, arising as it did out of a contract placed with the Boeing Aircraft Company for a surface-to-air missile system rather than specifically for electronic analogue computer equipment. However, in the late 1940s the newly independent US Air Force established and funded an R&D project at the Goodyear Aircraft Co. The principal aim of this project was to develop electronic analogue computer systems to be used in the study and design of aircraft and guided missile systems. The Goodyear project led to the GEDA range of commercial general-purpose electronic analogue computers.

In the late 1940s and 1950s analogue computer development was also undertaken at the US Air Force's think tank, the RAND Corporation. And in the mid-1950s the Air Force funded the development of a state-of-the-art analogue computer laboratory for the

Aeronautical Research Laboratory of the Wright Air Development Centre, at the Wright-Patterson Air Force Base in Ohio.

Boeing Aircraft Company: BEAC and BEMAC

In 1945 the US Army Air Force placed a contract with the Boeing Aircraft Company for the development of a surface-to-air missile system known as GAPA (Ground-to-Air Pilotless Aircraft).[47] Between 1945 and 1950 two general-purpose analogue computers were developed by staff at Boeing in connection with the GAPA programme. The first was a large electro-mechanical system used primarily as a simulator for missile and aeronautical research. This computer, known as BEMAC (Boeing Electro-Mechanical Analogue Computer), was developed entirely for in-house use and was never commercialised.[48]

The second of the two systems was called the BEAC (Boeing Electronic Analogue Computer). The BEAC was developed by Boeing's Physical Research Unit located in Seattle, Washington. BEAC was developed originally as an in-house design aid in connection with the GAPA missile project. However, other departments at Boeing soon began to borrow the original BEAC. It proved so useful that Boeing allocated funds for the Physical Research Unit to further develop and build improved BEAC systems for the firm's Acoustics and Electrical, Aerodynamics, Mechanical Equipment, Power Plant, and Structures departments. The subsequent commercial manufacture of BEAC began largely in order to provide the Physical Research Unit's developmental shop with regular work.[49] The BEAC became available commercially in 1950.[50] Though various production models of the BEAC were developed, Boeing did not develop a complete range of analogue computer products comparable with other contemporary manufacturers. Probably because of the ready availability of commercial systems that were larger and achieved much higher precision, Boeing discontinued the manufacture of their BEAC computer systems in the late 1950s.[51]

Goodyear Aircraft Corporation: GEDA

The Goodyear Aircraft Corporation—a subsidiary of the Goodyear Tire and Rubber Company—first became involved in the development of electronic analogue computers in 1947 with GEDA, the Goodyear Electronic Differential Analyzer.[52] GEDA was developed at Goodyear's Aerophysics Department, under a United States Air Force contract.[53] It was developed within a programme to construct a general-purpose DC electronic analogue computing facility at Goodyears's R&D and manufacturing headquarters at Akron, Ohio. This computing facility was known as the Dynamic Systems and Computations Laboratory (DSCL). By the mid-1950s the DSCL was similar in size to the largest of the laboratories at the ONR's Project Cyclone, the DSCL having 400 computing amplifiers versus Cyclone's 420. The majority of the computing equipment in the DSCL was owned by the US Government, but on bailment to Goodyear. Consequently, this meant that though it was available for use by private firms, external government contract work and in-house work for the US Air Force took precedence.[54]

The Goodyear Aircraft Corporation began manufacturing GEDA equipment on a commercial basis in 1949, and throughout the 1950s and early 1960s the development and manufacture of a wide range of GEDA computing equipment continued.[55]

The RAND analogue computer

Project RAND, established in 1945, was originally operated as a semi-autonomous division of the Douglas Aircraft Corporation. Engineers, physicists and mathematicians were recruited from Douglas, other aircraft companies and elsewhere. It was originally intended that Project RAND should be an advisory agency which could liaise between industry and the various services on the viability of new aircraft armaments and study the "techniques of air warfare". In November 1948 the RAND Corporation was established and redefined as an autonomous non-profit research institute, but remained *ipso facto* the newly independent US Air Force's main adviser on R&D. RAND's first "electronic brain" was an analogue computer—a Reeves REAC.[56]

The nature of the analogue-computer development work under-taken at RAND differed from that at either Goodyear or Boeing. Whereas Goodyear and Boeing developed their own electronic analogue computers virtually from scratch, RAND's involvement in analogue computer development began, like many others in industry and academia, with the purchase of an early-version REAC system. RAND bought their first REAC computer in 1948. It was used in a

FIGURE 4.2   THE RAND ANALOGUE COMPUTER[57]

great variety of applications. Many of these were associated with the study of theoretical aspects of the flight of aircraft and missiles. In 1951 A. S. Mengel of the RAND Corporation described the REAC as a "valuable tool in applied research". Yet problems in using the REAC led to a proposal to substantially modify the REAC. This was approved and a team of engineers, under the direction of William F. Gunning, set about rebuilding the REAC equipment. By 1951 almost every aspect of the original REAC had been redesigned and a larger and much more flexible system emerged, known as TRAC (The Rand Analogue Computer), see Figure 4.2.[58]

One important innovation which originated from RAND was the development of a removable patchboard system for the REAC. Early-model REACs had a patchcord and switchboard systems (similar to those used in manual telephone exchanges) to set up or program the machine. The main disadvantage of this method was that every time someone wanted to investigate a new problem they had to pull out the previous setup before they could set up their own pattern of connections.[59] Gunning came up with the idea of using an IBM plugboard to replace the telephone switchboard method. Using a removable plugboard meant that individual problems could be set up in advance, exchanged easily and stored for re-running later.[60] Removable plugboards similar to the RAND system were subsequently adopted by Reeves and other analogue computer manufacturers. From the early 1950s they were a standard feature on almost all general-purpose analogue computers.

The Aeronautical Research Laboratory, Wright Air Development Centre and the Systems Dynamics Synthesizer

In the late 1940s and early 1950s the Air Force also purchased commercial equipment and began to establish its own analogue computing centres. Two of the largest were the Holloman Air Development Centre (Computer Branch, Technical Analysis Division) located at Holloman Air Force Base, New Mexico, and the Aeronautical Research Laboratory of the Wright Air Development Centre at the Wright-Patterson Air Force Base, Ohio.[61] The computer installation at the Wright Air Development Centre is worth describing in more detail, since it was the first US Air Force-funded programme comparable with the earlier US Navy projects, in term of both its scale and overall influence on the future direction of electronic analogue computation. The growing use and large scale of the analogue computer development at the WADC reveals some of the contemporary attitudes among analogue computer users in the mid-1950s and suggests that a fairly high level of confidence persisted that analogue computer technology had a future quite distinct from the burgeoning digital computer domain.

In 1948, Colonel L. I. Davis, then Chief of the Armament Laboratory at the Wright-Patterson Air Force Base, was assigned the task of setting up and taking charge of a new Aeronautical Research Laboratory (ARL) at Wright-Patterson.[62] In the early 1950s the ARL became the major component of the newly formed Wright Air Development Centre (WADC). In the 1950s much of the work at the WADC was performed under the auspices of the Directorate of Weapons Systems, which was responsible for the development and procurement of new weapon systems for the US Air Force. This involved the evaluation of proposals for new weapon systems from firms competing for government contracts. The Directorate monitored the progress of projects after contracts had been placed and oversaw the testing of prototype systems.

The increasing technical complexity of new weapon systems compli-cated the task of evaluating and verifying their performance. This problem was especially acute for new weapon systems which, to operate effectively, relied on the successful integration of several sub-systems, including radar, navigation and guidance systems, launch and propulsion systems. Such difficulties highlighted inadequacies both in the manufacturer's and in the Air Force's ability to thoroughly check system performance. To overcome such inadequacies, computer simula-tion techniques were among those that the staff from the Computation Branch of the ARL, and other laboratories at the WADC, developed and used to evaluate complex weapon systems.

The first electronic analogue computer used by the Computation Branch of the ARL was a small special-purpose system borrowed from MIT's Draper Instrumentation Laboratory, where it had been developed for in-house use.[63] In 1949 the Computation Branch of the ARL bought its first Reeves REAC. By 1955 the laboratory con-tained four REACs and two Goodyear GEDA analogue computers. There was also a Bendix digital differential analyzer and an ERA 1103 digital computer.[64] By 1955, however, it had become evident to engineers and scientists at WADC that their existing analogue com-puting facility was too small for the computational tasks at hand and that a far larger computer facility was required. In contrast to the sit-uation in the mid-1940s, when the ONR sought to construct their large simulation laboratories, in 1955 there were several firms manu-facturing analogue computers commercially. This meant that the Air Force was able to define an initial specification, based on the type of problems to be solved, and then invite existing manufacturers to propose various solutions. L. Milton Warshawsky was the Project Scientist from the ARL's Computation Branch in charge of the pro-gramme to procure a new computer facility. Warshawsky and his staff combined what they considered the best aspects of the designs submitted by the private firms, with their own ideas for the new com-puter system. The result was a specification that not only consisted of the most recent developments in analogue computing but went beyond the current "state-of-the-art" commercial analogue computer systems.

Six private firms participated in this joint effort to derive a com-bined specification: RCA Laboratories, Princeton NJ, Electronic

Associates Inc., Long Branch NJ, Beckman Instruments, Richmond Ca, Mid-Century Instrumatic, New York NY, Goodyear Aircraft, Akron, Ohio, and the Reeves Instrument Corp., New York, NY. Once the specification was finalised, they were then invited to bid against each other for the contract to develop and build the computer facility. The Reeves Instrument Corp. won the competitive bid and was awarded the $1.5 million US Air Force contract on 1 June 1955, and the agreement stated that the computer was to be completed and accepted by 1 June 1956.[65]

The ARL's new computer was called the "Systems Dynamics Synthesizer" (SDS), but it also known as the "APACHE"—Automatic Programmed Analogue Computer Highly Expensive.[66] It was designed to be a general-purpose computing facility, allowing not only for its main purpose—"to aid in the analysis of air weapon systems"—but also to "satisfy the demand for the solution of a wide range of other engineering and mathematical problems of interest to the Air Force."[67]

Flexibility and ease of problem setup were major considerations in the design of the SDS. Set-up procedures were largely automated. Keyboards and punched paper tapes were used to set up coefficient pots, diode function generators and to run system checks before the real problem-solving computation began. Reliability and accuracy were two other key aspects of the design specification. Economic considerations undergirded these technical considerations: the SDS was to be used in the development and evaluation of new multi-million dollar weapon systems, and errors would be expensive.[68]

The design specification for the SDS was self-consciously an ambitious one.[69] Indeed, towards the end of 1955, it became evident that Reeves had underestimated the difficulties in implementing the ambitious design specification. Many of the off-the-shelf commercial computing units could not be used without first undergoing considerable development work to improve their accuracy, bandwidth and overall performance. Consequently, the project overran its schedule. The first of five computing "stations" was delivered on 1 June 1957, one year after the entire system was supposed to be up and running. The last shipment of equipment from Reeves did not arrive until the beginning of 1958, and at the end of the year staff from the ARL's Computation Branch were still debugging the system. More than a year and a half late, the ARL's Systems Dynamics Synthesizer cost the US Air Force approximately $3 million, twice the original estimate.[70]

Yet even before the WADC computer was completely installed, Reeves had incorporated aspects of the WADC system into their commercial product range, including a "Patchbay Verifier" originally developed by L. Milton Warshawsky of the WADC, and an "Auto-Control System" which permitted the computer to be set up via a keyboard or paper tape input; computer setups could also be recorded onto the paper tape and thus easily restored at a later date.[71] In the meantime, EAI and Beckman, both of whom had been involved in the early collaborative effort to design the WADC, had developed similar automated set-up and control system for their commercial analogue computer products.

The SDS remained in operation until the mid-1960s, when it was replaced by a large true-hybrid system consisting of a Comcor Ci 5000 analogue computer linked to a Scientific Data Systems' Sigma 7 digital computer.[72]

US ARMY PROJECTS AND ANALOGUE COMPUTER DEVELOPMENT 1945–1955

In the postwar years, the Army, Navy and Air Force carried out separate missile development programmes. Each service had its own outlook on the nature of future missile development. The US Army's own immediate postwar research and development in missile and rocket technology was dominated by experiments with the German V-2 rockets. Under Operation Paper Clip, German scientists, engineers, as well as complete rockets, spare parts, test equipment and so forth were transported from Peenemunde to the USA immediately after the end of World War II. The former staff and equipment from Peenemunde were taken to Fort Bliss in El Paso, Texas, and early in 1946 experiments with the V-2 rockets began. In preparation for the time when the supply of German V-2s would be exhausted, the US Army, in cooperation with the General Electric Company, established the Hermes Programme in 1946 to develop and manufacture new experimental rocket systems. The Hermes Programme was the origin of a series of short and medium-range surface-to-surface ballistic missiles—Corporal, Redstone, Jupiter, Sergeant and Pershing—developed for deployment by the Army.

Among the equipment and personnel transferred from Peenemunde to the USA were a special-purpose electronic analogue computer there during the war and its creator, Helmut Hoelzer (see chapter 3).[73] At Fort Bliss in the late 1940s Hoelzer's computer was employed in simulation studies of the US Army's Hermes rocket. In late 1949 the Army's missile programme was transferred from Fort Bliss to the Redstone Arsenal in Huntsville, Alabama, and the Guided Missile Development Division was established. This was the US Army's main ballistic missile research and development centre. At Huntsville, in 1950, the construction of a second, larger analogue computer was completed. This was based on and constructed in part from components of the original system designed by Hoelzer during the war. This computer was used in simulation studies of the Army's Redstone and Jupiter ballistic missiles and remained in service for ten years.[74]

The Army, research and development and indirect support for the electronic analogue computer industry

The US Army also used electronic analogue computers for the development of conventional weapons and other research activities. For example, in the 1950s the US Army Electronics Research and Development Laboratory (Oakhurst Field Station) at Fort Monmouth, New Jersey, operated a medium-sized analogue computer facility manufactured by George A. Philbrick Researches Inc. Problems associated with the effects of atmospheric electricity were among those investigated there.[75]

By 1955 there were two electronic analogue computers in use at the Army's Ballistics Research Labs at the Aberdeen Proving Ground: one in the Guidance and Control Branch consisting of approximately 30 operational amplifiers, and one in the Exterior Ballistics Laboratory consisting of 48 operational amplifiers. The Pitman-Dunn Laboratory at the Frankford Arsenal in Philadelphia was also among the Army-owned R&D centres that operated an electronic analogue computer facility. The first Frankford Arsenal analogue computer contained 100 operational amplifiers and was used in the study of ordnance problems associated with internal and external ballistics of various types of artillery.[76]

In the mid-1950s interior ballistic studies were also being carried out at the US Army's Picatinny Arsenal at Dover in New Jersey. The first electronic computer at Picatinny Arsenal was a Beckman 1000 series analogue computer. It was installed in the summer of 1954 at a cost of $88,000 and remained in service till 1959. Initially it was used in studies of the interior ballistics of closed breech guns. Demand for use of the analogue computing facility grew and soon outstripped the capacity of the Beckman system. In 1957 the Picatinny Arsenal purchased a PACE 31 R system costing $125,000 from Electronic Associates Inc. The PACE 31 R was in use throughout the 1960s and was not disposed of till 1973. However, during the 1960s the Picatinny Arsenal also hired hybrid computer equipment. This continued until 1969 when the newly formed Computer Systems and Evaluation Command in Washington DC gave approval for the purchased of a $1.2 million EAI 8900 hybrid computer system.[77]

The US Army also indirectly supported the electronic analogue computer industry through the R&D contracts it placed with private firms. In the late 1940s and 1950s several private firms working on the development of guided missiles and other weapon systems for the US Army developed and/or purchased and used electronic analogue computers. Among those that did were the General Electric Corp., which worked on the development of the Hermes ballistic missile, the American successor to the German V-2 rocket, the Western Electric Company/Bell Telephone Laboratories, which worked on the development of the Nike Ajax surface-to-air missile for the US Army,[78] and the Douglas Aircraft Company, which developed the Nike Hercules surface-to-air and Honest John surface-to-surface missiles.[79]

Following the explosion of a hydrogen bomb by the Soviet Union in 1953, all three armed services accelerated their ballistic missile development programmes. In 1956 the Guided Missile Development Division was superseded by the Army Ballistic Missile Agency (ABMA), and a joint programme with the US Navy to develop the Jupiter IRBM was established. Because of the high degree of duplication of R&D effort across the various missile programmes, the roles to be played by each of the three branches of the armed forces were redefined in 1956. The US Army was restricted to the development of short-to-medium-range surface-to-surface missiles, and responsibility for long-range ICBMs was given to the Air Force. Overall, the result was that the role of the US Army in ballistic missile development contracted, while that of the Air Force and the Navy—with its new Polaris missile programme—expanded.

Without a major role to play, the Army Ballistic Missile Agency languished until 1958 when the civilian National Aeronautics and Space Administration was created. In 1960 the more than four-and-a-half thousand employees of the ABMA, Huntsville, were transferred from US Army control to NASA and the facility renamed the George C. Marshall Space Flight Centre.[80] From the late 1950s NASA became one of, if not the, largest single purchaser and user of commercial analogue and hybrid computer equipment.

CONCLUSION

In the USA by 1950, military sponsored research and development had laid the technological and commercial foundations of a distinct postwar electronic analogue computer industry. Consequently, by the mid-1950s more than a half a dozen firms were manufacturing general-purpose electronic analogue computers, and the community of producers and users was expanding. In the following chapter I describe the structure and dynamics of the electronic analogue computer industry from 1945 through to the shift to hybrid computers in the 1960s, and the decline of the industry and electronic analogue and hybrid computing in the 1970s.

NOTES

1. R. E. Hare and W. E. Willison, "Analogue Computers: Part 1", *Instrument Practice*, Nov 1958, pp. 1200–1207, p. 1201.

2. K. Flamm, *Creating the Computer: Government, Industry, and High Technology*, The Brookings Institution, Washington, DC, 1988, p. 78

3. Each service had its own outlook on the nature of future missile development and there was much inter-service rivalry. The Army, for example, saw missiles as a form of long-range artillery. The Navy were primarily interested in missiles which could be launched from seagoing vessels. The Air Force's predilection for airplanes led to the development of relatively slow, winged, cruise-type missiles. See, W. von Braun and F. I. Ordway III, *History of Rocketry & Space Travel*, Nelson, London, 1967.

4. The ONR also funded digital computer projects, most notably Whirlwind at MIT, Hurricane (later called RAYDAC) at the Raytheon Corporation, CALDIC at the University of California and the Harvard Mark III. K. Flamm, *Creating the Computer: Government, Industry, and High Technology*, The Brookings Institution, Washington DC, 1988.

5. By 1948 about 40% of all basic research in the US was funded by the ONR. See K. C. Redmond and T. M. Smith, *Project Whirlwind: The History of a Pioneer Computer*, Digital Press, Bedford, Mass., 1980, p. 105; D. K. Allison, "US Navy Research and

Development since World War II", in M. R. Smith, *Military Enterprise and Technological Change: Perspectives on the American Experience*, MIT Press, Cambridge, Mass., 1985, pp. 289–328; J. L. Penick, *et al.*, (eds.), *The Politics of American Science, 1939 to the Present*, MIT Press, Cambridge, Mass., 1972, pp. 180–188; H. M. Sapolsky, "The Origins of the Office of Naval Research", in D. M. Masterson, (ed.), *Naval History: The Sixth Symposium of the US Naval Academy*, Wilmington, De., Scholarly Resources, 1987; D. Dickson, *The New Politics of Science*, Pantheon, New York, 1984.

6.   A. Karen and B. Loveman, "Large-Problem Solutions at Project Cyclone", *Instruments and Automation*, Vol. 29, Jan 1956, pp. 78–83. In 1952 responsibility for the project was transferred from the ONR to the US Navy's Bureau of Aeronautics; R. Mccoy, *et al.* "Designing an Office-Size Electronic Analog Computer", *Electrical Manufacturing*, Vol. 47, 1951, pp. 94–9, & pp, 230–2, & p. 234. It is worth noting that in late 1940s the Reeves Instrument Corporation decided to set up a digital computer group and hired Samuel Lubkin to head the operation. The machine that Reeves proposed to build was called the REEVAC and was based on the EDVAC design. However, before long Reeves decided not to proceed with the digital computer project, Samuel Lubkin left to join the digital computer group at the National Bureau of Standards and Reeves continued with its analogue computer development programme. Lubkin did not stay with the NBS for very long. Within a few months he left to form his own company the Electronic Computer Corporation which brought out a low cost digital computer called the ELECOM 100. See, S. Snyder, Smithsonian Computer History Project—Oral History Collection, National Museum of American History Archive Centre, Washington DC., Typescript of Interview, 16 July 1970, pp. 25–26.

7.   *Project Cyclone Symposium 1 on REAC Techniques*, Reeves Instrument Corporation Under Contract With Special Devices Centre, New York, March 15–16, 1951.

8.   S. Fifer, "Introduction", *Project Cyclone Symposium 1 on REAC Techniques*, Reeves Instrument Corporation Under Contract With Special Devices Centre, New York, March 15–16, 1951, p. 5.

9.   H. Zagar, "Description of the Reeves Electronic Analog Computer (REAC)", *Mathematical Tables and Other Aids to Computation*, Vol. 3, no. 24, Oct 1948, pp. 326–327; R. Mccoy, *et al.* "Designing an Office-Size Electronic Analog Computer"

10.   The 17 cabinets consisted of 14 servomechanism units and 3 cabinets of miscellaneous equipment. The oldest laboratory (number 3) contained older modified equipment, including 5 REACs, 4 servomechanism units and 2 special units, totalling 120 computing amplifiers. A. Karen and B. Loveman, "Large-Problem Solutions at Project Cyclone", p. 78.

11.   Ibid., pp. 80–81.

12.   Ibid., p. 78.

13.   S. Frost, "Compact Analog Computers", *Electronics*, July 1948, pp. 116–20, p. 117.

14.   *Project Cyclone Symposium 1 on REAC Techniques*, p. 2–4. Because of the close ties between electronic analogue computers and military weapons development programmes, those attending the symposium were required to obtain official clearance prior to their participation. A second Cyclone symposium was held in New York in 1952, during which the first day was reserved for the presentation of eight "classified papers". *Project Cyclone Symposium 2 on Simulation and Computing Techniques*, Reeves Instrument Corporation Under Sponsorship of the US Navy Special Devices Centre and the US Navy Bureau of Aeronautics, New York, April 29 to May 2, 1952.

15.   B. Dewey, (Vice President Reeves Inst.,) Letter to Professor William G. Shepard (University of Minnesota Institute of Technology), Minneapolis, 14 June 1949. CBI Archives, File: Marvin Stien 3–29.

16. M. Rees, "The Federal Computing Machine Program", *Science*, Vol. 112, Dec 1950, pp. 731–736; S. Fifer, "Introduction" *Project Cyclone Symposium 1 on REAC Techniques*, p. 5; E. P. Gray and J. W. Follin, Jr., "First Report on the REAC Computer", *Report TG-50*, The Applied Physics Laboratory, Johns Hopkins University, 15 Dec 1948.

17. *Project Typhoon Symposium 3, Simulation and Computing Techniques*, sponsored by Bureau of Aeronautics and US Naval Air Development Centre, Johnsville, Pa., Oct 12–14, 1953; *RCA an historical perspective*, RCA, Cherry Hill, NJ, 1978, pp. 14–15.

18. G. A. Korn and T. M. Korn, *Electronic Analogue Computers (D-c Analog Computers)*, 2ed., McGraw-Hill, New York, NY, 1956, p. 395.

19. E. A. Goldberg, "Stabilization of Wide-Band Direct-Current Amplifiers for Zero and Gain", *RCA Review*, June 1950 pp. 296–300.

20. R. R. Jenners, *Analog Computation and Simulation: Laboratory Approach*, Allyn and Bacon, Boston, 1965, p. 10. AC analogue computers continued to be used in certain areas of computation where the DC computer was considered an unproven technology; this was most in evidence in the field of special-purpose electronic analogue computers used in flight crew trainers. S. Fifer, *Analogue Computation: Theory, Techniques and Applications*, Vol. 3, McGraw-Hill, New York, NY, 1961, p. 618.

21. Anon., "New RCA Electronic Computer Aids US Air Defence; ...", RCA, New York NY, 21 Nov 1950, pp. 1–5, p. 1.

22. Ibid., pp. 1–2.

23. The first of these was the "trajectory model", in which two fluorescent balls suspended from thin metal rods provided a three-dimensional representation of the paths flown by the target (bomber) and radar-guided missile. At the same time, a scale model of the guided missile (thirty centimetres long by nine centimetres in diameter) rotated and pitched up or down, as it responded to the control signals that the full-size anti-aircraft missile would have received. Ibid.

24. Anon., "RCA Computer Aids Missile Design", *Product Engineering*, Jan 1951, p. 18.

25. Contract NOa(s)-10413 between the US Navy Bureau of Aeronautics and the Western Electric Company was authorised on 15 June 1949. Anon., *Navy Intercept Project Quarterly Engineering Report No. 18*, 1 Oct 1953–1 Jan 1954. pp. 1–2, AT&T Archives, Case 26656–100.

26. Simulations such as those performed on Typhoon were extremely complex. For example, one involved 56 equations, the equivalent to a 28th-order system. Anon., *An Intercept System Using All-Weather Rocket-Armed Interceptors in the Period 1956–1960*, Vol. 1 of 2, March 1, 1955. p. 57; Anon., *Navy Intercept Project Quarterly Engineering Report No. 18*, pp. 1–2; Anon., *Interceptor Terminal Positions and Headings from Simulated Vectoring Operations*, 29 Feb 1956. p. 1, AT&T Archives, Case 26656–100; Anon., *Automatic Carrier Landing Study: Final Report*, April 1954. AT&T Archives, Case 26656–89.

27. A. Hass, "Gigantic Computer Goes to Junk Pile", *Philadelphia Inquirer*, Philadelphia, Pa., 19 Sept 1968.

28. Archive Centre, David Sarnoff Laboratory, RCA, Princeton, NJ.

29. Ibid.

30. K. L. Wildes and N. A. Lindgren, *A Century of Electrical Engineering and Computer Science at MIT, 1882–1982*, MIT Press, Cambridge, Mass., 1985, pp 210–227, p. 227. Towards the end of the war a project to develop a universal flight trainer the ASCA (Airplane Stability and Control Analyser) was started at the Servomechanism Laboratory under the directorship of Jay W. Forrester. Originally intended to use analogue technology, Forrester turned instead to digital to implement the design. This project was transformed by Forrester into the famous Whirlwind digital computer project at MIT and later into the SAGE early warning system. Yet Edwards argues that the choice of digital was not neces-

sary taken on technical grounds. For a detailed account of the motivations and reasoning behind the re-definition and re-orientation of this project, see: P. N. Edwards, *The Closed World: Computers and the Politics of Discourse in Cold War America*, MIT Press, Massachusetts, 1996, chapter 3.

31.   A. C. Hall, "A Generalized Analogue Computer for Flight Simulation", *Trans.*, *AIEE*, Vol. 69, 1950, pp. 308–312, p. 308.

32.   Ibid., 308.

33.   Engineers at MIT found that the transformation and resolution of vector quantities arising from guided missile studies was a difficult operation for contemporary DC computer equipment to perform. They chose instead to employ a 400 Hz suppressed-carrier signal for data transmission. This system had the advantage of allowing the use of electromechanical vector resolvers and avoided the amplifier drift problems associated with early DC operational amplifiers. The choice of carrier signal meant that the relatively simple feedback amplifier could not be used for integration and so rate-servomechanisms were used. H. Jacobs, Jr., "Equipment Reliability as Applied to Analogue Computers," *Journal of the Association for Computing Machinery*, Vol. 1, Jan 1954, pp. 21–26.

34.   Approximately 16,000 solutions were computed in 1951, and in 1953 an estimated 20,000 solutions were obtained. W. W. Seifert and H. Jacobs, Jr., "Problems Encountered in the Operation of the M.I.T. Flight Simulator", 11 April 1952, *Typescript*, MIT Archives MC6, p. 1; W. W. Seifert and H. Jacobs, Jr., "The Role of Simulators in Relation to Low-Altitude Multiple-Target Problems", Paper presented at the US Redstone Arsenal, 5 Jan 1953, *Typescript*, MIT Archives MC6, p. 11.

35.   W. W. Seifert and H. Jacobs, Jr., "The Role of Simulators in Relation to Low-Altitude Multiple-Target Problems", p. 11; To provide some idea of the complexity of problems run on the DACL simulator, in 1953 in a three-dimensional trajectory problem the following mathematical operations were performed: 49 integrations, 130 multiplications of a constant by a variable, 99 multiplications of a variable by a variable, 6 divisions, 3 functions of two variables, 17 functions of one variable, 2 noise generators and "lots of additions". W. W. Seifert, "Seminar for Committee on Machine Methods of Computation", 27 Oct 1953, *Typescript*, MIT Archives MC6, p. 10.

36.   K. L. Wildes and N. A. Lindgren, *A Century of Electrical Engineering and Computer Science at MIT, 1882–1982*, p. 227.

37.   W. W. Seifert and H. Jacobs, Jr., "The Role of Simulators in Relation to Low-Altitude Multiple-Target Problems", p. 11; W. W. Seifert, "Seminar for Committee on Machine Methods of Computation".

38.   K. L. Wildes and N. A. Lindgren, *A Century of Electrical Engineering and Computer Science at MIT, 1882–1982*, p. 227.

39.   CURTIAC introduced several improvements in the design of multiplier servomechanisms and in the range of functions which could be derived from the primary variable: time. G. A. Korn and T. M. Korn, *Electronic Analogue Computers (D-C Analogue Computers)*, McGraw-Hill, New York, 1952, pp. 337–363.

40.   *Special Products News: The Beckman EASE Computer*, Beckman Instruments Inc., 18 April 1952; *EASE Computer*, Berkeley Division, Beckman Instruments Inc., c. 1954; Anon., *Electronic Analogue Computation Laboratory: Bulletin No. 2*, Department of Engineering, UCLA, Ca., 12 Dec 1950. The prototype systems developed at UCLA had 30 amplifiers which could be adapted to perform various mathematical functions in real-time.

41.   At first, Beckman manufactured EASE systems at their Special Products Department in South Pasadena, Ca. In July 1952 Beckman acquired the Berkeley Scientific Corp., based in Richmond, California. The corporation was renamed the Berkeley Division, Beckman Instruments Inc., and manufacture of EASE was transferred to the Richmond site. *Moody's Industrial Manual*, New York, 1954. The original EASE EC500 range cost between $4,500

and $5,500. This was soon superseded by the EASE 1000 Series, costing between $6,000 and $10,000, and subsequently by the EASE 1100 Series.

42. J. M. Carroll, "Analog Computers for the Engineer", *Electronics*, June 1956, pp. 122–129.

43. The NRL's REAC installation was small compared to Cyclone or Typhoon. It was therefore not possible to do large-scale missile simulations on it. Instead, it was used in a variety of applications as a general-purpose design and development tool by engineers and applied mathematicians. W. A. McCool, "Electronic Analog Computation", *Naval Research Laboratory Report*, Washington DC, December 1950.

44. At the Naval Air Missile Test Centre, the new REAC was first used on the Navy's Lark and Sparrow missile projects. The Lark missile project which began in 1944 was one of the first surface-to-air systems to be tested at NOTS, Inyokern, and later at the NAMTC, Point Magu. Test flights of the Sparrow air-to-air guided missile began at NAMTC, Point Magu in August 1948. D. S. Fahrney (R. Admiral), "Guided Missiles: US Navy The Pioneer", *American Aviation Society Journal*, Vol. 27, 1982, pp. 15–28.

45. The Puget Sound Naval Shipyard, Navy Bureau of Ships, bought a Boeing BEAC to use on problems in shipbuilding and design. Anon., *Computing Machine Service*, Bureau of Ships, Department of the Navy, Navyships 250-223-1, 1 June 1953.

46. *The Lightning Empiricist*, George A. Philbrick Researches, Boston, Mass., Vol. 1, no. 1, June 1952. By 1955 analogue computers were also being used by the Naval Gun Factory, (Physics Branch,724) Washington, DC, the Naval Ordnance Laboratory, Corona, Ca., the Naval Postgraduate School (Departments of Mathematics, Mechanics and Electrical Engineering), Monterey, Ca., and the Navy Electronics Laboratory, San Diego, Ca.

47. Over one hundred GAPAs were built before the project was cancelled in 1949. The GAPA programme was superseded by the Air Force's Bomarc surface-to-air missile programme between 1950 and 1951. See, W. von Braun and F. I. Ordway III, *History of Rocketry & Space Travel*, Nelson, London, 1967, p. 144; A. R. Weyl, *Guided Missiles*, Temple Press, London, 1949, pp. 109–111.

48. C. B. Crumb, Jr., "Engineering Uses of Analog Computing Machines", *Mechanical Engineering*, Vol. 74, 1952, pp. 635–639.

49. Anon., "The Boeing Electronic Analog Computer: The use of analog equipment in the study of dynamic systems", *Internal Typescript Report*, Physical Research Unit, Boeing Airplane Co., Seattle, c. 1950, p. 1.

50. BEAC was a low-cost ($7,000 in 1952) general-purpose DC amplifier analogue computer housed in a single two-metre-high cabinet. The BEAC was smaller than either the REAC or GEDA systems, having only twelve operational amplifiers. C. B. Crumb, Jr., "Engineering Uses of Analog Computing Machines".

51. The 1956 version, the BEAC 7000 ($4,433) was less expensive but very much resembled the original BEAC, though it had 20 rather than 12 operational amplifiers. J. M. Carroll, "Analogue Computers for the Engineer", pp. 124–125.

52. During World War Two the Goodyear Aircraft Corporation, a subsidiary of the Goodyear Tire and Rubber Company, operated a US government plant at Akron, Ohio. In 1947 the parent company purchased the factory in Akron from the US government for $2.2 million. From the late 1940s until the early 1960s this was the site where Goodyear designed, developed and manufactured its range of general-purpose electronic analogue computers.

53. M. Rees, "The Federal Computing Machine Programme", p. 731; *GEDA: Analog-computing equipment*, Trade publication, Goodyear Aircraft Corp. Akron, Ohio, 1954. The Goodyear GEDA had 20 DC operational amplifiers.

54. P. Hermann, *The Simulation Council Newsletter*, in *Instruments & Automation*, Vol. 28, Aug 1955, p. 1308.

55.	GEDA Computing Units	Price in 1954 ($)
	Main Computing Consols	13,500
	GN215-L3 Linear Analyzer	8,000
	GN215-N3 Non-Linear Analyzer	
	Auxiliary Equipment	1,250
	Electronic Multiplier(N3A)	1,250
	Electronic Multiplier(N3B)	1,675
	Servo-Multiplier(N3D)	1,500
	Function Generator(N3F)	1,000
	Summer Unit(N3Q)	1,275
	Reference Voltage(N3C)	200
	Decade Resistance Unit	500
	Potentiometer Unit(N3M)	825
	Impedance Bridge	975
	Two-Channel Record Amp	4,200
	Input-Output Table	6,200
	Six Channel Recorder	
	Total Cost	**$42,530**

Source: *GEDA Price List*, Goodyear Aircraft Corporation, Akron, Ohio, 1 March 1954; R.L. Hovious, C.D. Morrill and N.P. Tomlinson, "Industrial Uses of Analog Computers", *Instruments and Automation*, Vol. 28, April 1955, pp. 594–601.

56. W. F. Gunning, "A Survey of Automatic Computers, Analogue and Digital", *Report P 356*, The RAND Corporation, Santa Monica, Ca., 23 Dec 1953; W. F. Gunning, Smithsonian Computer History Project—Oral History Collection, National Museum of American History Archive Centre, Washington DC., Typescript of Interview, 9 Oct. 1972, pp. 8–11.

57. W. S. Melahn, "Modification of the RAND REAC", *Project Cyclone Symposium 1 on REAC Techniques*, pp. 41–44, p. 44.

58. W. S. Melahn, "Modification of the RAND REAC", *Project Cyclone Symposium 1 on REAC Techniques*, pp. 41–44.

59. A. S. Mengel and W. S. Melahn, "RAND REAC Operators' Manual", *Report RM-525*, Project RAND, Santa Monica, Ca., 1 Dec 1950.

60. This, however, was before the Consent Act of 1956, and at the time IBM refused to sell not only machines but also spare parts. Eventually the Air Force Chief of Staff intervened and persuaded the President of IBM, Thomas J. Watson Jr., that IBM should sell RAND the plugboards they required. P. Armer, Smithsonian Computer History Project—Oral History Collection, National Museum of American History Archive Centre, Washington DC., Typescript of Interview, 17 April 1973, pp. 14–15.

61. L. B. Wadel and A. W. Wortham, "A Survey of Electronic Analog Computer Installations", *Trans., IRE, Electronic Computers*, Vol. EC-4, Pt. 2, June 1955, pp. 52–55.

62. One of the formative influences on the development of analogue computing at the ARL was the work undertaken at Columbia University during the war. Colonel Davis showed prospective staff of the new laboratory the paper by J. R. Ragazzini *et al.*, (1947), to gauge their interest in working there. L. M. Warshawsky, "My Career in Simulation", in J. McLeod (ed.), *Pioneers and Peers*, The Society for Computer Simulation International, San Diego, Ca., 1988, pp. 24–27.

63. In the late 1940s and early 1950s staff at the Draper Lab developed several special-purpose compressed-time-scale electronic analogue computers to simulate and solve differential equations associated with the Lab's R&D work on navigation and guidance systems for aircraft and missiles. Samuel Giser and Frank S. Spada, the founders of the GPS (General Purpose Simulation) Instrument Company, which manufactured high-speed repetitive-operation analogue computers were amongst those working on the development of analogue computers at the Draper Lab in the late 1940s and early 1950s. P. A. Holst, "Sam Giser and the GPS Instrument Company: Pioneering compressed-time scale (high-speed) analog computation", in J. McLeod (ed.), *Pioneers and Peers*, Society for Computer Simulation International, San Diego, Ca., 1988, pp. 4–10.

64. In 1954 the WADC analogue computer installation consisted of four REAC C101s/S101s and two GEDA L3s/N3s, in total it contained approximately 130 amplifiers. The Computation Branch also had a OARAC digital computer manufactured by General Electric and an IBM CPC (replaced in mid-1955 by an ERA 1103). L. M. Warshawsky, "My Career in Simulation", p. 24; L. M. Warshawsky, "Letter to the Editor", *The Simulation Council Newsletter*, The Simulation Council, Camarillo, Ca., March 1954, pp. 1–8, pp. 5–6.

65. *The Wall Street Journal*, New York, Monday, August 22, 1955, p. 15.

66. APACHE occupied an air-conditioned room approximately 52 by 27 feet. Four identical computing stations were located in each corner. A fifth unit, the central control console, was located midway between two of the corner units. It contained 160 integrating amplifiers, 196 summing amplifiers, 144 inverters, plus 50 servo-multipliers, 24 electronic multipliers, 32 ten segment diode function generators and 16 servo-resolvers. L. M. Warshawsky, "WADC's New Large Analog Computer", *Proc., First Flight Simulation Symposium*, Nov 15–16, 1956, pp. 247–256; L. M. Warshawsky, *The Simulation Council Newsletter*, in *Instruments & Automation*, Vol. 31, Nov 1958, pp. 1848–1849.

67. L. M. Warshawsky, "WADC's New Large Analog Computer", p. 247–248.

68. L. M. Warshawsky, *The Simulation Council Newsletter*, in *Instruments & Automation*, Vol. 32, Dec 1959, pp. 1861–1864.

69. Throughout the 1950s there was considerable debate over the relative merits of analogue and digital computing techniques. By 1955, electronic analogue computers had according to L. M. Warshawsky "proved their right to exist" but, because of the rapid developments in all aspects of digital computing many saw this as a transitory condition. The Systems Dynamics Synthesizer design was thus seen as a rejoinder from the advocates of analogue computing, stating that it had not only a past but also a future. L. Milton Warshawsky, who supervised the design and development of the ARL's SDS commented in 1955:

> It will be apparent to many of you that what we have done, in general, has been to make convenient many of the test procedures that you and we have been using for the past several years. The clever baling wire tricks that have developed among the many users of this type of equipment become awkward, burdensome and uneconomical when applied to a "colossus electronicus." They were justifiable, perhaps, during the phase when analog computers were trying to prove their right to exist. This machine is our way of expressing our confidence that electronic analog computers have an important role to play and we hope that we have helped this infant science to grow up.

L. M. Warshawsky, "WADC's New Large Analog Computer", p. 256.

70. L. M. Warshawsky, *The Simulation Council Newsletter*, p. 1848–1849.

71. H. E. Harris, "New Techniques for Analogue Computation," *Instruments & Automation*, Vol. 30, May 1957, pp. 894–899.

72. L. M. Warshawsky, "My Career in Simulation", p. 25.

73. J. E. Tomayko, "Helmut Hoelzer's Fully Electronic Analogue Computer", *Annals of the History of Computing*, Vol. 7, no. 3, July 1985, pp. 227–240.

74. H. H. Hosenthien and J. Boehm, "Flight Simulation of Rockets and Spacecraft, in E. Stuhlinger, F. I. Ordway, III, J. C. McCall and G. C. Bucher, (eds.), *From Peenemunde to Outer Space*, NASA George C. Marshal Space Flight Centre, Huntsville, 1962, pp. 437–469.

75. *The Lightning Empiricist*, George A. Philbrick Researches Inc, Boston, Mass., Vol. 11, no 3, Jan 1963, p. 4.

76. Anon., *The Simulation Council Newsletter*, in *Instruments & Automation*, Vol. 31, Feb 1955, pp. 5–6; L. B. Wadel and A. W. Wortham, "A Survey of Electronic Analog Computer Installations".

77. A. G. Edwards, "It began with computers", in J. McLeod, (ed.), *Pioneers and Peers*, The Society for Computer Simulation International, San Diego, 1988, pp. 62–65. The EAI systems at the US Army's Picatinny Arsenal remained in operation until 1980. In 1977 an EAI 7900 hybrid was purchased at a cost of $500,000 and it too remained in service until 1980. In 1978 a EAI HYSHARE 700 System costing $400,000 was installed and remained in service till 1987.

78. In their development of the Nike missile system, Western Electric used the general-purpose analogue computer, known as Gypsy, developed by the Bell Telephone Laboratories between 1945 and 1949. W. E. Ingerson, "The Nike-Ajax computer", *Bell Laboratories Record*, Vol. 38, no. 1, 1960, pp. 26–30; J. L. Troe, "Nike in the air defence of our country", *Bell Laboratories Record*, Vol. 38, no. 4, 1960, pp. 122–129.

79. Prior to 1952 the Douglas Aircraft Corporation did not have its own electronic analogue computer facility and instead relied upon the use of outside equipment at BTL, Project Cyclone, Project Typhoon, the Dynamic Analysis and Control Laboratory at MIT and the Navy Air Missile Test Centre, Point Magu, Ca. At the end of 1951 Douglas bought its first Reeves REAC and a Network Analyzer manufactured by the William Miller Co. In the 1950s the use of analogue computing equipment within the Douglas Aircraft Company continued to grow. By the mid-1950s the firm had a large electronic analogue computer at its plant in El Segundo, Ca: the analogue computer contained 286 operational amplifiers, plus auxillary computing and data-recording equipment. C. R. Strang, *Computing Machines in Aircraft Engineering*, Douglas Aircraft Company, Santa Monica, Ca., 12 Dec 1951.

80. W. von Braun and F. I. Ordway III, *History of Rocketry & Space Travel*, p. 165.

CHAPTER 5

Commercialisation, Hybridisation and Competition: The Electronic and Hybrid Computer Industry in the USA, 1945–1975

INTRODUCTION

One of the chief aims of this book is to establish that there was indeed a distinct post-war electronic analogue computer industry and community of users, one which from the late 1940s to the mid-1960s grew steadily, levelled out, and in the 1970s declined.

In this chapter, I describe the development and growth of the manufacturing and user base for electronic analogue and hybrid (combined analogue/digital) computers in the USA, from 1945 to the mid-1970s. I begin by introducing a classification scheme to analyze and describe, largely at the level of the firm, the origins, structure and dynamics of the nascent electronic analogue computer industry.[1] This covers the period from 1945 to 1955, by which time a mature industry had emerged. Indeed, as we shall see, by the mid-1950s more than a dozen firms were manufacturing a large range of general-purpose electronic analogue computer systems. These varied in size from desk-top models to room-sized installations, and in cost from $10,000 to more than $200,000. As a result of both in-house development of electronic analogue computers and the availability of commercial systems, the community of users also expanded considerably during this period. From 1955 to the mid-1960s the manufacturing and user base continued to grow, and firms began to introduce hybrid computer systems into their range of products. As we shall see, hybrid computers were a "user-led" development, initiated in the mid-1950s by aeronautics firms in connection with the development of intercontinental ballistic missiles (ICBM). From the mid-1960s, hybrid computing provided the dominant context for the continued development of electronic analogue computers.

OVERVIEW OF THE ELECTRONIC ANALOGUE COMPUTING SCENE, 1945–1955:
DEVELOPERS, MANUFACTURERS AND USERS

By 1950 a manufacturing base consisting of four firms—Reeves, Boeing, Goodyear and G. A. Philbrick Researches Inc.—had been established, along with a market for general-purpose electronic analogue computers. Analysing the formative period, 1945 to 1955, we see that the firms and institutions involved in electronic analogue computer development can be divided into three categories: developer/users, user/manufacturers and non-user/manufacturers. These are defined as follows:

— The developer/users are those firms and institutions that developed and built general-purpose analogue computer systems for in-house use, which were never commercialised. Example are the Dynamic Analysis and Control Lab at MIT, Bell Telephone Labs, Sperry Gyroscope Co,[2] and Convair Inc.
— The user/manufacturer group consists of those firms that developed and used analogue computers in-house, but also made the transition to the commercial manufacture of electronic analogue computers. Examples are the Boeing Aircraft Co., the Goodyear Aircraft Co., and the Reeves Instrument Co.
— The non-user/manufacturer group consists of those firms that developed and built electronic analogue computers not primarily for in-house use, but as a commercial products. Some of these firms existed already and diversified, while others were newly established to develop analogue computer systems primarily as commercial products. Examples are GAP/R Inc, Radio Corporation of America, Berkeley Division of Beckman Instruments, EAI, GPS Instruments Inc, Mid-Century Instrumatic Inc and the Donner Scientific Co.

In chapter 4 we have seen that military projects and funding played a crucial role in the development of postwar electronic analogue computing. Table 5.1 gives an overview of their origins and indicates the

Table 5.1 Electronic Analogue Computers: Origins and Classification, 1945–1955

Category	Principal Sponsors			
	US Navy	USAF/USAAF	US Army	Non-Specific
Developer/user	Curtiss-Wright CURTIAC; DACL, MIT Flight Simulator	RAND Corp. The Rand Analogue Computer	H. Hoelzer AC analogue computer/ rocket simulator	Bell Telephone Labs GPAC/GYPSY; Sperry Gyroscope Sperry Diff Analyzer; Convair Inc, Pomona and San Diego, Large GPEACs
User/ manufacturer	Reeves REAC	Boeing BEAC;		
Non-user/ manufacturer	Project Cyclone RCA Project Typhoon; Beckman, EASE	Goodyear GEDA		George A. Philbrick Researches, K2 and K3 units; Electronic Associates Inc. EAI 16–24 D; Donner Scientific Donner 30; Mid-Century, MC400; GPS Inst. Co, GPS-6

TABLE 5.2 FIRMS MANUFACTURING ELECTRONIC ANALOGUE COMPUTERS, 1945–1955

First wave firms	Second wave firms	
Firms manufacturing analogue computers prior to 1950	Major firms that began manufacturing analogue computers between 1950 and 1955	Minor firms that began manufacturing analogue computers between 1950 and 1955
Boeing Airplane Company, Seattle, Washington	Beckman Instruments Inc, (Berkeley Division), Richmond, Calif.	Computer Corporation of America, New York, NY
Goodyear Aircraft Company, Akron, Ohio	Donner Scientific Company, Berkeley, Calif.	Dynamic Analysis Inc, Van Nuys, Calif.
George A. Philbrick Researches Inc, Boston, Mass.	Electronic Associates Incorporated, Long Branch, NJ	Heath Company, Benton Harbour, Mich.
Reeves Instrument Corp, New York, NY	GPS Instrument Company, Boston, Ma.	Weber Aircraft Co, Burbank, Calif.
Radio Corp of America, Princeton, NJ	Mid-Century Instrumatic Corp, New York, NY	

relations between military sponsorship and the categories of developers, users and manufacturers defined above.

By the early 1950s the technological basis and demand for electronic analogue computers was established, and a "second wave" of manufacturing firms emerged. The majority of these firms were from the electronics industry and were already manufacturing instrumentation and communication systems, and subsequently diversified into the commercial manufacture of electronic analogue computers. Examples are Beckman Instruments Inc. (Berkeley Division), Electronic Associates Inc., the Donner Scientific Company, and Mid-Century Instrumatic Inc. Also one private firm was established solely to develop and manufacture electronic analogue computers and associated equipment: the GPS Instrument Company, see Table 5.2.

These new non-user/manufacturer firms formed the core of the specialised analogue computer manufacturing base and increasingly dominated the commercial manufacture of analogue and subsequently hybrid computer systems. By 1955 nine major and a number of smaller firms were manufacturing analogue computing equipment. Of these, only Boeing and Goodyear started as developer/users and made the transition to user/manufacturers. By the mid-1950s, per annum sales of general-purpose electronic analogue computers amounted to several million dollars. Reported sales in 1954 totalled approximately $6 million, and in 1955 the figure rose to $10 million.[3] General-purpose electronic analogue computers ranged in size from desk-top units having between 10 and 24 operational amplifiers to large room-sized installations with 200, or more, operational amplifiers. Prices varied greatly, from $1000 for a 10-operational-amplifier machine to a large room-size system like the Beckman EASE 1200 which, when fully expanded, cost in the region of $250,000. Table 5.3 gives details of the commercial electronic analogue computers available between 1948 and the mid-1950s, and their manufacturers.

COMPANY PROFILES: MANUFACTURERS OF GENERAL-PURPOSE ELECTRONIC ANALOGUE COMPUTERS, ESTABLISHED BEFORE 1956

G. A. Philbrick and repetitive-operation electronic analogue computers

When the war ended, George A. Philbrick became a consultant engineer for the Foxboro Co. (see chapter 3) and made plans to enter

TABLE 5.3 ELECTRONIC ANALOGUE COMPUTERS, MANUFACTURERS AND MODELS, 1948 TO THE MID-1950S

Manufacturer	Date Introduced	Computer Models	Description	Price
Beckman Instruments Inc, Berkeley	1952	EASE EC500	Small (10 op-amp) desk-top general-purpose electronic analogue computer.	$5,000
Division Richmond, Calif.	1954	EASE 1031	40 op-amp GPEAC	$30–65,000
		EASE 1032	80 op-amp GPEAC, with central control panel	$100–$200,000
	1956	EASE 1132	60 op-amp GPEACs with optional digitally controlled input/output system	$100,000–$250,000
	1956	EASE 1133	100 op-amp version	
	1956	EASE 1200	Large central console controlled GPEAC with digitally controlled input/output system	$250,000
Boeing Airplane Company, Seattle Washington	1949 to 1956	BEAC to BEAC 7000	Series of GPEACs, solve up to 12th-order differential equations	$7,000–$4,500
Donner Scientific Company, Berkeley Calif.	1955	Donner 30	Small desk-top general-purpose electronic analogue computer.	$1,000
Electronic Associates Incorporated, Long Branch, NJ	1952	EAI 16–24 A	One-off general-purpose electronic analogue computer for the Westinghouse Corp.	$—
	1953	EAI 16–24 D	Large dual patch-board GPEAC.	$20,000
	1953	PACE 16–31 R	Range of expandable high precision GPEAC equipment	
			24 op-amp system	$19,500
			120 op-amp system	$85,400

Manufacturer	Date Introduced	Computer Models	Description	Price
Goodyear Aircraft Company, Akron, Ohio	1949	GEDA	GPEACs, solve up to 12th-order differential equations	$13,000
	1951	GEDA GN215 L2		$13,500
	1952	GEDA GN215 L3	Linear GPEAC, solve up to 12th-order differential equations	
			Non-Linear GPEAC	$8,000
	1953	GEDA GN215 N3	Combined L3 & N3 system	$43,000
	1956	GEDA A-14		
GPS Instrument Co, Boston, Ma	1952	General Purpose Simulator GPS-6	Compressed time scale computer for NACA, Moffett Field	$13,000
	1955	GPS-12B	Expandable repetitive operation/ compressed time scale GPEAC, solves 6th order differential equations: the GPS-12B was similar but had twice the computing capacity.	$15,000–$25,000
	1952	MC 400		
Mid-Century Instrumatic Corp, New York, NY	1954		MC 500Small portable desk-side GPEAC	$5,000
George A. Philbrick Researches Inc, Boston, Mass.	1948	K2 and K3 Series	30 op-amp expandable GPEAC system	$10,000 up
			Plug-in repetitive-operation electronic analogue computing devices; expandable to form computers of various sizes	$500–$100,000

TABLE 5.3 ELECTRONIC ANALOGUE COMPUTERS, MANUFACTURERS AND MODELS, 1948 TO THE MID-1950S (CONT'D)

Manufacturer	Date Introduced	Computer Models	Description	Price
Reeves Instrument Corp, New York, NY	1948	REAC	18 op-amp expandable GPEAC	$—
	1949	REAC C101	20 op-amp expandable GPEAC	$14,000 up
	1951	REAC 200	20 op-amp expandable GPEAC	$15,000 up
	1953	REAC 300	20 op-amp expandable GPEAC	$15,000 up
	1955	REAC C301/ C302	12 op-amp desk top GPEAC	$5–10,000
	1955	REAC 400	20 op-amp expandable GPEAC with optional digitally controlled input/output.	$20,000– $1,000,000
Radio Corp of America, Princeton, NJ		Project Typhoon	Large one off guided missile simulator	$1,500,000
Computer Corporation of America, New York, NY	1952	IDA ERS-2	Small desk-top GPEAC	$7,000–$8,000
Dynamic Analysis Inc, Van Nuys, Calif.	1955	LDA-3	Small desk top GPEAC	$3,500
Heath Company, Benton Harbour, Mich.	1954	ES 100	Small GPEAC in kit form	$500–$1000
Weber Aircraft Co, Burbank, Calif.	1956		General-purpose analogue computer system	$8,000 up

MIT as a graduate student. His intention was to gain a doctorate, using the design of a novel high-speed electronic analogue computer as a thesis project. In 1946, while involved in the preliminary stages of this project, he was approached by the Wright Aeronautical Corporation (later Curtiss-Wright) to build a special-purpose turbo-prop engine control simulator. Philbrick postponed his graduate studies and accepted the $22,000 contract. Philbrick formed his own company in November 1946; he named the company George A. Philbrick Researches Incorporated, or (GAP/R) for short, and began to build the engine control simulator in the bedroom of his own house in Cambridge, Massachusetts. Shortly after the simulator for the Wright Corporation was completed, Philbrick received an order from the National Advisory Committee on Aeronautics for a similar system.[4]

Philbrick never returned to his doctoral project at MIT. Instead, he developed a range of small, modular electronic analogue computing devices known as K3s. The K3 series of "little black boxes" could be interconnected to model control processes and to solve differential equations and other mathematical problems. Each K3 unit was identified by a single letter, associated with the function of the unit: A, B, C, H, J, L, Z.[5] Philbrick's analogue computer products differed in two ways from the commercial systems which dominated the analogue computer market in the late 1940s and early 1950s. Whereas other major manufacturers sold complete analogue computing systems, GAP/R manufactured, marketed and sold the K3 series as individual components. Small and large systems could then be built up by interconnecting as many "little black boxes" as necessary.[6] The other major difference was that Philbrick's analogue computing equipment operated on the repetitive-operation (rep-op) basis.

In the postwar years, two broad categories of analogue computing technology emerged. One was the single-shot and the other the repetitive-operation computer. In a single-shot computer the problem is set up and run, and a single solution is plotted on paper. Parameters may then be varied and the problem re-run. On Philbrick's repetitive computing equipment a problem could be automatically re-run at between 10 and 60 times per second, and the solution displayed on an oscilloscope. Parameter variations could be made during the run and the effects observed almost immediately.[7]

However, until 1955 DC electronic analogue computers which operated on the single-shot principle, such as the REAC, GEDA, BEAC, EASE and EAI's PACE computers, dominated the general-purpose analogue computer market, gaining widespread acceptance among engineers. From the mid-1950s, repetitive-operation analogue computing started to receive more attention.[8] The development of higher-performance rep-op computer systems by GPS Instruments Inc. and later by Computer Systems Inc. (to aid in the development of ICBMs) was the main spur to the growth of interest in rep-op analogue computing. By the early 1960s, all of the major manufacturers of single-shot equipment had redesigned their computers to enable calculations to be run in either single-shot or rep-op mode.

GPS Instrument Company

The GPS (General Purpose Simulator) Instrument Company was founded in 1952 by Samuel Giser and Frank S. Spada. At the time, Giser and Spada were working at the Draper Instrumentation Laboratory at MIT. In the Draper Laboratory, Giser worked on the design and development of special-purpose electronic simulators. These were used in simulation studies of the airframe and in inertial navigation control systems under investigation in the Laboratory. He realised that a single general-purpose computer could be designed, that would significantly reduce the repetitive and wasteful engineering effort incurred in building individual simulators for each new design proposal. Giser and his staff subsequently designed and built a repetitive operation electronic analogue computer.

In 1952 the National Advisory Committee on Aeronautics (NACA) expressed an interest in purchasing one of the Draper Laboratory computers built by Giser and his staff. NACA intended it for use at their Ames Aeronautical Laboratory at Moffett Field, California. When others at MIT showed little interest in building the computer system for NACA, Giser decided to obtain permission from MIT to take on the $13,000 contract himself. Giser and Spada established the GPS Instruments Inc. and built the computer outside of normal working hours at Giser's home in Sharon, Massachusetts. For the next three years, Giser and Spada continued to run GPS Instruments Inc. on a part-time basis and built and sold several similar computing systems. In

1955 they decided to leave MIT in order to run their company on a full-time basis.[9]

In the mid-1950s the only other firm manufacturing compressed-time-scale, repetitive-operation analogue computers on a commercial basis was George A. Philbrick Researches Inc. Yet the rep-op equipment manufactured by GAP/R Inc was relatively slow and imprecise compared with the computers developed by GPS Instruments, which operated on a compressed time scale of 3000:1 compared with the 1000:1 ratio used by GAP/R Inc. GPS's computer systems were, however, considerably more expensive than those of its main rival. GPS continued to develop and manufacture high-speed analogue computer systems until the early 1970s.

Electronic Associates Incorporated

Electronic Associates Incorporated (EAI) was established in October 1945 by Lloyd F. Christianson, Arthur L. Adamson and nine other associates from the US Army Signal Corps. The firm began as a consultancy and design group, specialising in shipboard and aircraft communications systems. In 1952 EAI sold its first general-purpose analogue computer to Westinghouse, where it was used in design studies of the reactor for the US Navy's first nuclear-powered submarine, the Nautilus. Soon afterwards, EAI launched their first of a range of standard computing units: the Model 16–24 D. Their computer products were a great commercial success, and EAI's net sales jumped from approximately $1,070,000 in 1952 to $4,274,000 in 1953. In the years that followed, EAI became the world's leading manufacturer of high-precision analogue and hybrid computer equipment.[10]

Beckman Instruments Incorporated (Berkeley Division)

Beckman Instruments Inc. diversified into the manufacture of electronic analogue computers in 1952. Its first analogue computer system, the Beckman EASE (Electronic Analogue Simulating Equipment), had its origins in a collaborative project between the US Navy's Ordnance Test Station (NOTS) based at Inyokern, Chinalake, California, and staff from the University of California at Los Angeles. The original EASE was based on a computing system, of the same name, developed jointly by Dr Tom Rogers, his Ph.D student Louis G.

Walters at UCLA, and engineers from NOTS. Initially, it was developed to aid in rocket research and in the simulation and testing of ordnance equipment.[11]

Beckman Instruments commercialised and began to manufacture the EASE computer system at its Special Products Department in South Pasadena, California. In order to establish a full-scale development and manufacturing facility for analogue computer products, Beckman Instruments acquired the Berkeley Scientific Corporation of Richmond, California. Thus, in July 1952 the Berkeley Scientific Corporation became the Berkeley Division, Beckman Instruments Inc., and the research, development and manufacture of Beckman's EASE analogue computer equipment was transferred to the Richmond site.[12] The original EASE EC500 range, which was demonstrated for the first time in May 1952, cost between $4,500 and $5,500.[13] By the mid-1950s Beckman had developed a wide range of general-purpose analogue computing equipment. In 1956 Beckman launched the EASE 1200 Series and consolidated its position as one of the leading manufacturers of analogue computers in the USA.[14]

Mid-Century Instrumatic Co., Computer Systems Inc. and Dian Laboratories Inc.

Mid-Century Instrumatic began to develop their range of analogue computers in 1952. Mid-Century's chief computer engineer was Urbano Manfredi. Between 1946 and 1952 Manfredi worked for the Reeves Instrument Corp. He was Head of the Design and Maintenance Section at the ONR's Project Cyclone, before joining Mid-Century. In 1955 Manfredi and the former Head of Project Cyclone, Stanley Fifer, established an analogue computing and consultancy centre known as Dian Laboratories. Based at Mid-Century's headquarters on Broadway in New York, the centre was also equipped with analogue computing equipment manufactured by Mid-Century. In 1959 Dian Laboratories broke off its association with Mid-Century and began to manufacture analogue computers systems on its own. The same year, Dian Laboratories launched the Dian 60, 80 and 120 analogue computer products.[15] Also in 1959, Mid-Century was restructured and became a subsidiary of the Schlumberger Group. Mid-Century Instrumatic became Computer Systems Inc., which in the 1960s developed and manufactured a range of high-speed, repetitive-operation and iterative analogue computer systems.[16]

Donner Scientific Co. and Systron-Donner Corp.

The Donner Scientific Company, of Berkeley, California, began to manufacture electronic analogue computer systems in 1955, with the Donner 30. By 1957 the firm had moved to new, larger premises in Concord, California. In 1959, following a merger with the Systron Corporation, the Donner Scientific Company became a subsidiary of the newly created Systron-Donner Corporation. Throughout the 1960s Systron-Donner manufactured analogue and subsequently hybrid computer products.[17]

Other manufacturers of electronic analogue computers, 1945 to the mid-1950s

Reeves, Boeing and Goodyear also began manufacturing electronic analogue computers before 1956; their entry into the industry has already been described in chapter 4. There were also a number of generally smaller firms producing commercial systems. In the early 1950s, the Computer Corporation of America, based in New York, manufactured a 20-operational-amplifier analogue computer system known as the IDA (Integro-Differential Analyzer) which sold for approximately $8,000. In the mid-1950s The Heath Company, of Benton Harbor in Michigan, a subsidiary of Daystrom Corporation, sold a 15-operational-amplifier computer in kit form for home construction, and Weber Aircraft of Burbank in California, was manufacturing a 24-operational-amplifier computer system which sold for approximately $8,000.

ELECTRONIC ANALOGUE COMPUTER INSTALLATIONS AND USERS, 1945–1955: A SURVEY

In 1954, a survey of analogue computer installations was undertaken by L. B. Wadel and A. W. Wortham of the Chance Vought Aircraft Inc. in Dallas, Texas. From this we know that there were at least 96 electronic analogue computer installations in the USA at that time.[18] However, many of the firms asked to respond did not do so, and major installations such as the ONR's Typhoon did not appear in the survey. Nevertheless, from this survey alone we can see that the

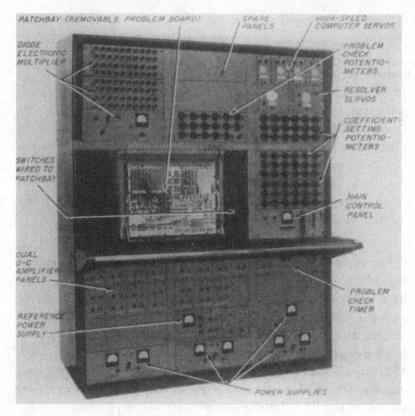

FIGURE 5.1   REEVES INSTRUMENT CO., REAC

increasing availability of commercial analogue computer systems in the early-to-mid 1950s was followed by a corresponding rise in the number of analogue computer installations, 50% of the total analogue computer installations between 1946 and 1955 being established in the last two-and-a-half years.

· Wadel and Wortham's data show that installations ranged in size from the smallest with 10 operational amplifiers to the largest with more than 560. Also, from their data we can calculate that prior to 1955, 45% of those computer installations listed were located in government-owned research facilities, firms, college and university departments directly involved in aeronautical research and development. Given the rapid growth of the aerospace industry on the West Coast of the USA in the postwar years, it is perhaps not surprising to

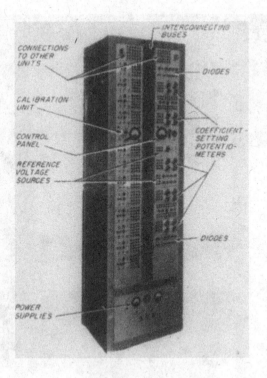

FIGURE 5.2 BOEING AIRCRAFT CO., BEAC

find that the Pacific region had more analogue computing installations than any other. Based on the number of operational amplifiers per installation, we find that by 1955 almost 38% of the total analogue computing capacity was located in the Pacific region.[19] The region with the second-largest analogue computing capacity was the Middle Atlantic region with 23%, see Table 5.4.[20]

Overall, the industrial users were the largest group, having more than twice the number of analogue computer installations than either the government facilities, or the academic institutions. Generally speaking, the industrial installations were also larger than those in either government or academic analogue computer facilities. In terms of the number of operational amplifiers, this meant that there was approximately five times the computing capacity in industry, compared with government facilities, and three times the computing capacity in industry compared with universities and colleges, see

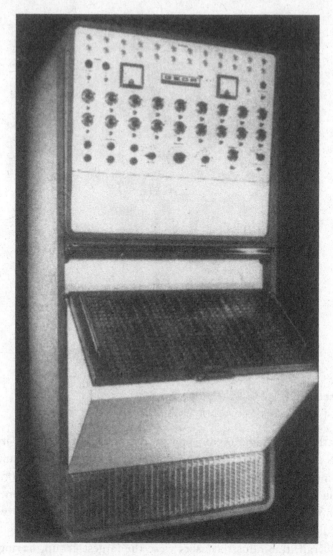

FIGURE 5.3   GOODYEAR AIRCRAFT CO., GEDA

Table 5.5.[21] Table 5.6 lists the analogue computer users who appeared in the Wadel and Wortham survey of 1955 and gives an indication of the size of the installations.

Characterising analogue computer development and use in the years between 1945 and 1955, we find that aeronautical applications

FIGURE 5.4   ELECTRONIC ASSOCIATES INC., EAI 16-31R

FIGURE 5.5   BECKMAN INST., CO., EASE

Figure 5.6   Donner Scientific Co., Donner 30

Table 5.4 Distribution of electronic analogue computer installations in the USA by region

Region	Number of Installations	Number of Amplifiers
Pacific	26	2981
Middle Atlantic	22	1831
East North Central	11	669
South Atlantic	6	404
Mountain	5	584
West North Central	5	398
West South Central	5	398
New England	4	516
East South Central	2	50

Table 5.5 Distribution of computing capacity (amplifers) in the USA by sector

User	Number of Amplifiers
Industrial	5060
Government	1627
Universities	1179

TABLE 5.6 ELECTRONIC ANALOGUE COMPUTER USERS, 1954

User	Installation Size/Access
Aerojet-general Corp, Azusa Ca	3N
Argonne National Lab, Lamont, Ill	3AG
Armour Research Foundation, Chicago, Ill	3A
Battelle Memorial Inst., Columbus, Oh	2A
Beckman Instruments Inc, Richmond, Ca	3A
Ballistics Research Lab,	
Aberdeen Proving Grd., Md	
Guidance and Control Branch	2N
Exterior Ballistics Lab	3N
Beech Aircraft Corp, Wichita, Kan	3N
Bell Aircraft Corp, Buffalo, NY	2A, 5AG
Canadair Ltd, Montreal, Canada	2N
CDC Control Services, Hatborough, Pa	2A
Chance Vought Aircraft, Inc, Dallas, Tex	5N
Collins Radio Co, Cedar Rapids, Ia	3N
Convair, Fort Worth, Tex	4N
Convair, San Diego, Ca	6AG
Convair, Pomona, Ca	6AG
Defence Research Lab, Austin, Tex	2AG
Detroit Arsenal, Centre Line, Mich	4AG
Douglas Aircraft Co Inc, El Segundo, Ca	1N
Douglas Aircraft Co Inc, Santa Monica, Ca	4N
Dow Chemical Co, Midland, Mich	1N
Dynalysis Develmt. Labs Inc, Los Angeles	4A
Electronic Associates Inc, Princeton, NJ	4A
Frankford Arsenal (Pitman-Dunn Lab),	
Philadelphia, Pa	3AG
General Motors Corp, Aeroproducts	
Operations, Vandalia, Ohio	3A
Gilfillan Brothers, Inc, Los Angeles, Ca	3N
Glen L. Martin, Co, Baltimore, Md	3N
Grumman Aircraft Engineering Corp,	
Bethpage, LI	4N
Holloman Air Development Centre, Computer	
Branch, Technical Analysis Div.,	
Holloman Air Force Base, NMex	5N
Hughes Aircraft Co, Culver City, Ca	
Guided Missile R&D Div	4N
Radar R&D Div	5N
Hughes Tool Co, Aircraft Div,	
Culver City, Ca	1N
Jack and Heintz, Inc, Pasadena, Ca	2N

TABLE 5.6 ELECTRONIC ANALOGUE COMPUTER USERS, 1954 CONT'D

User	Installation Size/Access
Jet Propulsion Lab, Pasadena, Ca	3N
Johns Hopkins Univ, Washington DC	4N
Leeds and Northrop Co, Philadelphia, Pa	1N
Lockheed Aircraft Corp, Burbank, Ca	4N
Louisiana State Univ	
(Elec Eng Dept), Baton Rouge,	2N
MIT, Dynamic Analysis and Control Lab,	
Cambridge, Mass	6AG
W. L. Maxson, Corp, New York, NY	2A
Minneapolis-Honeywell Regulator Co,	
Aeronautical Div, Minneapolis, Minn.	5N
NACA Ames Aeronautical Lab,	
Moffett Field, Ca	5N
NACA Langley Aeronautical Lab,	
Langley Field, Va	3AG
Naval Research Lab, Washington, DC	3AG
New York Univ, Coll of Eng, New York, NY	1A
North America Aviation Inc, Downey, Ca	4N
Northrop Aircraft Inc, Hawthorne, Ca	5N
Northwestern Univ, Aerial	
Measurement Lab, Evanston, Ill	3AG
Oregon State College, Mech Eng,	
Corvallis, Ore	1A
Picatinny Arsenal, Dover, NJ	
ORDBB-TR1	1AG
ORDBB-TH1	1AG
Polytechnic Inst of Brooklyn, Brooklyn, NJ	1A
Project Cyclone, Reeves Instrument Corp,	
New York, NY	6AG
Puget Sound Naval Shipyard, Bremerton, Wash.	2N
Purdue Univ, Div of Eng Sciences,	
West Lafayette, Ind	2N
Purdue Univ, School of Aeronautics,	
West Lafayette, Ind	2N
RCA, Radar Eng, Moorestown, NJ	1N
Ramo-Wooldridge Corp, Los Angeles, Ca	5N
RAND Corp, Santa Monica, Ca	3N
J. B. Rea Co, Inc, Santa Monica, Ca	4N
Redstone Arsenal, Computation Lab,	
Huntsville, Ala	2N
Rensselaer Polytechnic Inst,	
Computer Lab, Troy, NY	2A

TABLE 5.6 ELECTRONIC ANALOGUE COMPUTER USERS, 1954 CONT'D

User	Installation Size/Access
Republic Aviation Co, Guided Missiles, Hicksville, LI	2N
Sandia Corp, Sandia Base, Albuquerque, NMex	3N
Schlumberger Co, Res Labs, Ridgefield, Conn	2N
Southern Research Inst, Birmingham, Ala	1A
Sperry Corp, Great Neck, NJ	
Aeronautical Eng	4N, 4N, 3N
Surface Armament Div	3N
Taylor Model Basin, Navy Dept, Washington, DC	2N
Technical Operations Inc, Arlington, Mass	2A
Temco Aircraft Corp, Dallas, Tex	2N
Univ of Buffalo, Physics Dept, Buffalo, NY	1A
Univ of California, Electrical Eng, Berkeley, Ca	1A
Univ of Colorado, Eng Experiment Station, Boulder, Colo	1A
Univ of Kansas, Elect Eng, Lawrence, Kan	1A
Univ of Michigan, Willow Run Res Centre, Eng Res Inst, Ypsilanti, Mich	5A
Univ of Minnesota, Coll of Eng, Inst of Tech, Minneapolis, Minn	2A
US Navy Missile Test Centre, Simulation Lab, Point Mugu, Ca	4AG
US Naval Gun Factory, Physics Branch, Washington DC	1AG
US Naval Ordnance Lab, Corona, Ca	5A, 3N
US Naval Postgraduate School, Monterey, Ca	
Dept of Mathematics and Mechanics	1AG
Dept of Electrical Engineering	1N
US Navy Electronics Lab, San Diego, Ca	1AG, 1N
Westinghouse Electric Corp,	
Control Eng Dept, Buffalo, NY	1N
Aviation Eng Dept, Lima, Ohio	1N
Atomic Power Div, Pittsburgh, Pa	3N
Analytical Section	5L–51
East Pittsburgh	4N
Air Arm Div, Baltimore, Md	4N
White Sands Proving Ground, NMex	3N, 5N
Worcester Polytechnic Inst	
Elec Eng, Worcester, Mass	1A

TABLE 5.6 ELECTRONIC ANALOGUE COMPUTER USERS, 1954 CONT'D

User	Installation Size/Access
Wright Air Development Centre, Aeronautical Research Lab, Wright-Patterson USAF Base, Ohio	4AG

Key	Number of Operational amplifiers	Key	
1:	10–20	A:	Available to outsiders.
2:	21–40	AG:	Available to outside government contractors
3:	41–80	N:	Not available to outsiders
4:	81–160		
5:	161–320		
6:	321–640		

predominated. Moreover, the dominance of government projects, so evident in the early stages of the development and use of analogue computers, diminished as the technology diffused into industry. Indeed, industry rapidly took over as the primary source of new developments and as the largest user of analogue computers.

However, though aeronautics-related research and development was, and remained, the dominant influence, general-purpose electronic analogue computers were suitable for, and used in, a wide variety of other applications. For example, analogue computers were also widely used in nuclear engineering. From a survey performed in the USA in 1957, we find that of 28 firms and organisations using computers in nuclear engineering applications, 14 were using analogue computer systems, while all were using digital computers. Two out of the twenty analogue computers listed were constructed in-house, the rest were commercial systems.[22] Analogue computers were also employed increasingly in the chemical, automotive, electrical and electronics industries, and in universities and technical colleges. Table 5.7 gives a list of some analogue computer applications in the mid-1950s.

THE ELECTRONIC ANALOGUE COMPUTER INDUSTRY, 1955–1960

From the mid-1950s, the increasing availability of commercial computer systems meant that industrial research and development

Table 5.7 Partial list of electronic analogue computer applications to 1960

Aeronautics, Aircraft and Guided Missile Design

Flight simulation of aircraft and missiles
Missile trajectory computations
Rocket burning and flight analysis
Aircraft controls and auto pilot design
Aircraft blind landing systems
Turbojet and supercharger controllers
Helicopter vibrations and control analysis
Design of wind-tunnel mounting systems
Supersonic flow problem analysis

Electrical and Electronic Engineering

Servo-auto follower for radar tracking
Induction motor simulation
Design of frequency modulation transmission
Linear and nonlinear circuit design
Electron optics analysis
Feedback amplifiers and servo-control systems
Simulation and analysis of transmission lines

Nuclear Engineering

Dynamics of heavy water reactor
Reactor Fluctuation Kinetics
Reactor Load following characteristics
Coolant flow analysis
Power station control optimization

Mechanical Engineering Design

Ship and submarine control systems
Hydraulic control and transmission system design
Motor car suspension systems analysis
Vibrations of structures analysis
Dynamics of mechanical linkages analysis
Heat flow problem analysis

Chemical Engineering

Distillation column dynamic response
Chemical reactor batch reactor simulation
Chemical process control, design and analysis
Nuclear products process control

Miscellaneous

Bio-Medical simulation
Economic model simulation
Marine engineering analysis
Civil engineering design

laboratories, universities and military research centres tended to purchase rather than develop their own in-house analogue computer system. Electronic analogue computer development was increasingly dominated by the specialised firms. Competition between firms increased as the specialised manufacturers carved out a place in this relatively small market. This competition, combined with the demands from users, led to considerable improvements in the performance of DC operational amplifiers, electronic multipliers and function generators, and almost all technical aspects of electronic analogue computer performance.

In the mid-to-late 1950s, several changes took place in the industry. Overall, the industry continued to expand, and sales of electronic analogue computer equipment grew. One of the new firms to emerge in the late 1950s was Applied Dynamics Inc., of Ann Arbor, Michigan. Applied Dynamics was established by three professors from the University of Michigan and a mechanical design engineer, and in the 1960s grew to be one of the largest and most successful analogue/hybrid computer firms in the world. Towards the end of the 1950s, Mid-Century Instrumatic was taken over by Computer Systems Inc. At the same time, Dian Laboratories, which was Mid-Century's computing centre and consultancy service, became an independent company and began to manufacture its own range of electronic analogue computers.

In 1956, Boeing's period of diversification into the development and manufacture of analogue computers ended, and production of its BEAC computer was discontinued. Though no longer directly involved in the manufacture of analogue computers, Boeing became one of the analogue computer industry's largest customers. Indeed, the following year the Seattle Division of the Boeing Aircraft Company bought a large analogue computer system consisting of six Beckman EASE 1133 computer consoles, totalling some 600 operational amplifiers, at a cost in excess of $750,000.[23] By 1961 Boeing had three Beckman EASE analogue computer installations. Two were at its Aero-Space Division in Seattle, where they were used in the development of the US Air Force's Dyna-Soar, Minuteman ICBM and Bomarc programmes. The other belonged to Boeing's Transport Division, where it was used in connection with the design of the Boeing 707 and 720 airliners.[24]

In the late 1950s, many firms with older equipment began to modernise and/or expand their analogue computer facilities. So too did many of the research establishments that were under Federal control. For example, in 1957 Project Cyclone was modernised and expanded with systems from the new REAC 400 series.[25] In the mid-1950s, Beckman and EAI were two of the most successful of the new firms. Beckman won large orders from the Lockheed Aircraft Corp. One of the largest orders was for a computer installation for the company's Missiles and Space Division in Sunnyvale, California. This was used in the development of the POLARIS Fleet Ballistic Missile for the US Navy. Lockheed's Georgia Division, where the C-130 Hercules was developed and manufactured, also bought Beckman EASE equipment in the mid-1950s.[26]

Yet, though Beckman had considerable success with its growing range of analogue computer products, it was EAI that in the late 1950s began to establish a dominant position among the major manufacturers of electronic analogue computers: not only in the USA but also in Europe. Indeed, between 1952, the year the EAI delivered its first general-purpose electronic analogue computer, and 1959, net sales rose from just over $1 million to almost $14.5 million. By 1965 this figure had more than doubled, to $32.6 million, of which 70% was due to the sale of analogue computer systems.

By the mid-to-late 1950s, every major aircraft and aerospace company and government-owned aircraft and missile development and test facility in the USA owned or were hiring commercial electronic analogue computer systems. This computing capacity was supplemented by commercial computing centres, which almost all of the major analogue computer manufacturers operated, see Table 5.8. These computing centres generally contained the firm's latest computing equipment, which was for hire along with the expertise of the centre's staff. In part, these computing centres were showcases, but there was no shortage of demand, and a great deal of serious research and development work was undertaken in them. Generally, these computer centres were used by organisations that did not have equipment of their own. However, they were also employed by firms and military agencies that needed to supplement their existing computing capacity.[27] Table 5.9 lists the major electronic analogue computer manufacturers in the USA from the mid-1950s to 1960.

TABLE 5.8 ELECTRONIC ANALOGUE COMPUTING CENTRES ESTABLISHED BY AMERICAN FIRMS: 1950–1960

Firm	Location of Computing Centre	Opening Date
Beckman Instruments Inc, Berkeley Division Richmond, Calif.	Richmond, California Los Angeles, California	1955 1957
Electronic Associates Incorporated, Long Branch, NJ	Princeton, NJ Los Angeles, California Brussels, Belgium Burgess Hill, Sussex, England	1954 1956 1957 1960
GPS Instrument Co, Newton, Ma	Newton, Massachusetts	1958
Mid-Century Instrumatic/DIAN Laboratories Corp, New York, NY George A. Philbrick Researches Inc, Boston, Mass.	New York, NY Boston, Massachusetts	1955 1950

TABLE 5.9 ELECTRONIC ANALOGUE COMPUTERS, MANUFACTURERS AND MODELS, MID-1950S TO 1960

Manufacturer	Date Introduced	Computer Models	Description	Price
Applied Dynamics Inc. Ann Arbor, MI.	1957–1960	AD-1-16	8 op-amp desk top GPEAC	$2,800
		AD-1-32	32 op-amp GPEAC	$13,000
		AD-1-32PB	32 high precision op-amp GPEAC	$15,000
		AD-1-64PB	64 high precision op-amp GPEAC	$29,600
		AD-1-128PB	128 high precision op-amp GPEAC	$60,000
Beckman Instruments Inc,	1956	EASE 1132	60 op-amp GPEACs with	$100,000–
Berkeley Division Richmond, Calif.			optional digitally controlled	$250,000
			input/output system	
	1956	EASE 1133	100 op-amp version	
	1956	EASE 1200	Large central console controlled	$250,000
			GPEAC, with digitally controlled	
			input/output system	
Computer Systems Inc, New York, NY.	1959	MC 5800	Large repetitive operation/	$20,000–
(Formerly Mid-Century Instrumatic)			iterative GPEAC, expandable	$100.000
			to 134 op-amp system	
DIAN Laboratories Inc New York, NY	1959	DIAN 60	Range of GPEAC computer	$—
(Former computing consultancy service		DIAN 120	systems	
for Mid-Century Instrumatic)		DIAN 180		
Donner Scientific Company, Berkely Calif.	1957	Donner 3000	Small desk top general-purpose	$1,000–
			electronic analogue computers	$10,000
	1958	Donner 3400		

TABLE 5.9 ELECTRONIC ANALOGUE COMPUTERS, MANUFACTURERS AND MODELS, MID-1950S TO 1960 (CONT'D)

Manufacturer	Date Introduced	Computer Models	Description	Price
Electronic Associates Incorporated, Long Branch, NJ	1957	PACE 16-131 R	Expandable GPEAC with optional ADIOS (Automatic Digital Input/OutputSystem)	$20,000–$200,000
	1958	PACE 231 R	Range of medium to large GPEACs, single-shot, or high-speed repetitive operation, and ADIOS option	$20,000–$1,000,000
	1959	PACE TR-10	Desk-top fully transistorized GPEAC, with 10 volt operation, solves 10th order differential equations	$4,000–$11,000
Goodyear Aircraft Company, Akron, Ohio	1956	GEDA A-14	Large GPEAC, combining GEDA GN215 L3 and GEDA GN215 N3 analogue computer units.	$43,000
GPS Instrument Co, Newton, Ma	1958	GPS Statistical Analogue Computer	Expandable repetitive operation GPEAC, with statistical data analysis capability, operates on a 3000:1 compressed time scale.	$20,000–$200,000
Mid-Century Instrumatic Corp, New York, NY	1952 / 1954	MC 400 MC 500	Small portable desk-side GPEAC 30 op-amp expandable GPEAC system	$5,000 $10,000 up
George A. Philbrick Researches Inc, Boston, Mass.	1958	K5, K7 Series	Plug-in repetitive-operation electronic analogue computing devices; expandable to form computers of various sizes	$350–$220,000
Reeves Instrument Corp, New York, NY	1955	REAC 400	20 op-amp expandable GPEAC with optional digitally controlled input/output system	$20,000–$1,000,000

THE ICBM AND THE ORIGINS OF COMBINED ANALOGUE-DIGITAL OR HYBRID COMPUTING

In the early 1950s, comparisons of analogue and digital computing techniques led some to the view that a unified analogue/digital approach could be used to overcome the relative weaknesses of each computing method. The intended result was a computer with the high computing speeds of the electronic analogue computer and the high precision of the digital computer. However, in the early 1950s there was little real demand for combined analogue-digital equipment, since most of the problems in engineering and aeronautical design at that time could be solved satisfactorily using separate analogue or digital computers.[28] Besisde the lack of a clear application for such a complicated computing system, the costs and technical difficulties involved in developing a combined analogue-digital solution were a major disincentive to pursuing the idea further. However, while conventional computational or simulation techniques were adequate for most problems, the complexity, hazards and exceptional costs involved in ICBM development emphasised the inadequacies of existing partial simulation methods. The rapid acceleration of the US's ICBM programme in 1954 provided not only the technical imperative but also the funds to implement a combined analogue-digital system.

The first combined analogue-digital computer systems emerged from the US Air Force's Atlas ICBM programme. Attempts to simulate the Atlas missile presented particular difficulties because the designers wished to include the autopilot and other parts of the missile's hardware in a total-system simulation. This meant that the complete missile and control system simulation had to be performed in real time. Yet another problem was the study of the effects of random disturbances or "noise" on the operation of the missile system. This required a great many simulation runs, so that the probability of the missile hitting its target could be determined by statistical analysis of the data. All-digital simulation was too slow to deal with the high frequencies in some of the control loops, including those in the navigation system. All-analogue computing was fast, but not precise enough for trajectory calculation over the long flight path of the missile. This application suggested a computer system that combined extremely high-speed computing with high precision, data storage and analysis capabilities.

The perceived solution was the combination of a large general-purpose electronic analogue computer with a large scientific digital computer. These combination systems were later generally referred to as true hybrids. In the absence of commercially available alternatives, the true-hybrid computer became a user-led development. Existing commercial analogue computer manufacturers were in fact still busy establishing and improving their electronic analogue computer products. In the meantime, the aerospace industry was establishing the direction that commercial electronic analogue computer development would later take.

The first firms to design and actually implement combined analogue-digital systems were the Convair Division of the General Dynamics Corporation and the newly formed Ramo-Wooldridge Corporation. These systems were developed between 1954 and 1956 as a direct result of the computation and simulation requirement created by the Atlas missile project.

The Convair Division of the General Dynamics Corporation

The use of electronic analogue computers at Convair (formerly Consolidated Vultee) in San Diego, California, began in 1948. That year the US Navy supplied the company with a Reeves REAC, because it was one of several firms working on aspects of the Navy's LARK missile project. Problems with the early-model REAC computer led Stanley Rogers and a number of other engineers in the Dynamics Group at Convair to undertake the design of a new electronic analogue computing system. In the hope of being successful with only one, four proposals for funds to build this computer systems were submitted to government agencies. However, all four proposals were successful, and Convair was granted almost $3 million to build four electronic analogue computing systems. In 1950, Stanley Rogers was placed in charge of the group of design engineers working on the electronic analogue computer development project. Completed in 1953, three of the analogue computer systems were installed side by side at Convair's plant in San Diego, and the fourth went to Convair's Pomona site. The San Diego facility was one of the largest in the USA at that time, and consisted of more than 700 operational amplifiers.[29]

Convair's involvement in the Atlas missile project began in April 1946 with a contract from the USAAF to perform feasibility studies of the MX-774 Upper Air Test Vehicle. However, severe economies in military R&D resulted in the cancellation of the MX-774 project in June 1947. Nevertheless, Convair decided to continue the MX-774 project, using their own funds, and subsequently built and launched three MX-774 test missiles.[30] The climate created by the onset of the Korean War, and a growing belief that the USSR was actively pursuing ICBM research, led to a partial revision of US Air Force policy on ballistic missile development. This reappraisal, combined with encouraging results from test-firings and rocket research undertaken by Convair, led the US Air Force in 1951 to establish Project Atlas (or MX-1593), the successor to the MX-774 project. In February 1954 a complete revision of the Air Force's attitude towards ballistic missiles development followed the wholesale endorsement of an accelerated ICBM programme by the influential Strategic Missile Evaluation Committee. The possibility of attack by Soviet nuclear missiles redefined the priority of ICBM research, and funding restrictions which had been a major consideration in the years following the end of World War II, were swept aside. In 1955 the Air Force's Atlas and Thor missile programmes were assigned the highest of national priorities.[31]

Work at Convair on the construction of a combined analogue/digital computer began in the summer of 1954. The first stage involved the design of a signal-conversion unit to physically connect Convair's electronic analogue and digital computers (a Univac 1103A, replaced in 1955 by an IBM 704). This conversion equipment, known as the ADDAVERTER (Analogue-Digital-Digital-Analogue conVERTER), was developed for Convair by the Epsco Company of Boston, Massachusetts. The ADDAVERTER was delivered in September 1956 and cost approximately $200,000. By the end of the year the first combined analogue/digital simulation studies of the Atlas missile were under way.[32]

The Ramo-Wooldridge Corporation

The rapid expansion and massive scale of the Atlas missile programme led directly to the creation of the Ramo-Wooldridge Corporation of Los Angeles, California. In 1954 the US Air Force realised that an organisation was required to provide the technical management that such a complex and large-scale project demanded.

TABLE 5.10 TRUE-HYBRID COMPUTER INSTALLATIONS IN THE USA 1954 TO 1962

User/developer	Analogue	Digital	Interface	Application
Autonetics, Armament & Flight Control Div.	EAI PACE 231 R	Bendix G-15	Packard-Bell Multiverter	Complex guided weapon control system design
Convair, San Diego	In-house mainly EAI equipment	Univac 1103A & IBM 704	Epsco ADDAVERTA	Atlas missile simulation
Douglas Aircraft	EAI PACE 131 R	Bendix G-15	—	Aircraft and space systems simulation
General Electric, Missile and Space Vehicles Dept	EAI PACE 231 R	IBM 704	EAI HYCOL	Space navigation and control
Grumman Aircraft	Reeves REAC	IBM 704/7090	Adage Inc. data link	Chemical process simulation
IBM Research Lab Yorktown Heights	EAI PACE 231 R	IBM 704	EAI Add-a-Link	Navigation system simulation
IBM Research Lab Owego	EAI PACE 231 R	IBM 704	EAI Add-a-Link	
McDonnell Aircraft Corp. St Louis	EAI PACE 131	IBM 7094	In-house	Control system design
National Bureau of Standards	Mid-Century	NBS's SEAC	in-house	Ground controlled intercept systems
North America Aviation/ Naval Ordnance Lab, Corona	EAI PACE 231 R	Packard Bell TRICE	—	Missile simulation
Ramo-Wooldridge	EAI PACE 131 R	Univac 1103A	Epsco ADDAVERTA	Atlas missile simulation
United Aircraft	Beckman EASE 2133	DEC PDP 1	Packard Bell	V/STOL aircraft, helicopters and space systems simulation
University of Minnesota	Reeves REAC	ERA 1103	EAI Add-a-Link	Control systems

Ramo-Wooldridge was established initially to advise the Strategic Missile Evaluation Committee, but by the end of 1954 had assumed responsibility for the technical management of the entire Atlas missile programme. Under the direction of Simon Ramo and Dean E. Wooldridge, the corporation grew rapidly from 170 employees in 1954 to more than 5,000 in 1960. In 1957 the corporation merged to form the Thompson-Ramo-Wooldridge, or TRW, Corporation[33]

In 1955 Ramo-Wooldridge established an Analogue Computing Centre at its headquarters in Los Angeles. The analogue computing equipment for the Centre was purchased from Electronics Associates Inc., and consisted of three EAI PACE computer consoles containing a total of 300 operational amplifiers. Ramo-Wooldridge also ran a separate Digital Computer Centre, based on a Univac Scientific Model 1103A. Ramo-Wooldridge, like Convair, contracted the Epsco Company of Boston to build an ADDAVERTER to combine its large electronic analogue and digital computer installations. The Epsco signal-conversion unit was delivered in October 1956, and after a short commissioning period a programme of hybrid computer simulation studies of the Atlas missile began.[34]

Subsequently, in the late 1950s, several other firms and organisations experimented with hybrid computing. Most of these used analogue/digital computer interface equipment designed by EAI and Packard-Bell. The National Bureau of Standards in 1957 linked its electronic analogue computer, manufactured by Mid-Century Instrumatic, to its SEAC digital computer to solve problems associated with ground control intercept systems. The Missile and Space Vehicle Department at General Electric connected its EAI PACE analogue computer facility to an IBM 7070 to explore problems associated with space navigation and control. By the early 1960s there were more than a dozen true-hybrid computer installations in the USA, see Table 5.10.[35]

DEVELOPMENTS IN COMMERCIAL ELECTRONIC ANALOGUE AND HYBRID COMPUTING, 1960–1970

Until the mid-1960s, hybrid computing remained largely the province of a group of generally large user firms and organisations with both analogue and digital computer installations. The first commercial hybrid computers were much less ambitious in their design than the large true-hybrid systems already mentioned. Among the firms in the electronic analogue computer industry, the hybridisation of analogue with digital computing technology began rather tentatively. Initially it followed two distinct directions. The first of these involved the introduction of paper tape and keyboard input/output systems, to partially automate the time-consuming set-up and checking procedures associated with large electronic analogue computers.[36] The other main development was the incorporation of electronic digital logic circuits and components into the analogue computer. These were used to store results, to perform arithmetic calculations and to control the sequence of a series of analogue computations through conditional branching. The various blocks of digital logic, memory, arithmetic and conversion functions were set up through patchboards and patch cords similar to those used to interconnect the electronic analogue computer components (see chapter 3).

In this way, commercial electronic analogue computer systems incrementally tended towards true-hybrid computers. Yet even the EAI HYDAC 2000 which EAI launched in 1962, claiming that it was the first fully integrated hybrid computer, did not contain a general-purpose digital computer.[37] The major breakthrough both with the partial and incremental hybridisation of analogue computers and with thermionic valve technology came in 1964 with the introduction of the Comcor/Astrodata Ci 5000. The Ci 5000 was the first all-transistor electronic analogue computer that was designed specifically so that it could be completely controlled by a general-purpose digital computer. It was not long before the other major manufacturers had their own all-transistor, fully hybrid-compatible analogue computers on the market.[38]

Prior to the launch of the Ci 5000, several of the commercial manufacturers were already working on the development of transistorised hybrid computers. In the case of EAI, the firm had been spending an increasing proportion of its R&D budget on developing digital products since 1959, and by 1964 the company was almost ready to launch a new range of analogue/hybrid computers.[39] In 1964 EAI introduced the EAI 8400, a scientific digital computer designed for real-time all-digital or hybrid simulation. The following year EAI introduced its replacement for the extremely successful PACE 231-R series: the all-transistor EAI 8800 hybrid computer. The 8800

TABLE 5.11 Electronic Analogue Computers, Manufacturers and Models, 1960 to 1970[40]

Manufacturer	Date Introduced	Computer Models	Description	Price
Applied Dynamics Inc. Ann Arbor, MI.	1957–1960	AD-1-16	8 op-amp desk top GPEAC	$2,800
		AD-1-32	32 op-amp GPEAC	$13,000
		AD-1-32PB	32 high precision op-amp GPEAC	$15,000
		AD-1-64PB	64 high precision op-amp GPEAC	$29,600
		AD-1-128PB	128 high precision op-amp GPEAC	$60,000
	1962	AD-24-PB	24 op-amp high precision desk top GPEAC	$9,000
		AD-2-64PB	64 op-amp high precision GPEAC, PBC model	$—
		AD-2-64PBC	includes central console control facility.	$—
	1964	AD 256	Large fully transistorised GPEAC with digital facilities.	$—
	1966	AD 4	Fully transistorised GPEAC with digital facilities and integral hybrid interface.	$15,000– $140,000
	1969	AD 5	Fully transistorised, 10 volt, GPEAC computer with digital facilities and integral hybrid interface	$50,000– $450,000
Beckman Instruments Inc, Berkely Division Richmond, Calif.	1961	EASE 2100 Series	Range of medium to large GPEACs, with central console control, iterative and digital facilities and DO/IT system.	$30,000– $300,000
	1964	EASE 2200 Series	Range of large GPEACs, with central console control, and integrated hybrid interface.	$50,000 up
	1964	Beckman/SDS Hybrid Series	Range of hybrid computers, combining EASE 2200 analogue computer range with digital computers manufactured by Scientific Data Systems Inc.	$100,000– $500,000
		2200/SDS 920 2200/SDS 930		

Table 5.11 Electronic Analogue Computers, Manufacturers and Models, 1960 to 1970[40] (cont'd)

Manufacturer	Date Introduced	Computer Models	Description	Price
Computer Products Inc, South Belmer, NJ	1964	CP MkIII	Medium to large GPEAC with digital facilities and hybrid interface	$—
	1966	CP 10/50	Desk top, fully transistorised 10 volt GPEAC, . with digital facilities	$—
Comcor Inc, Denver, Colo. (a subsidiary of Astrodata, Anaheim, Calif.	1961	Ci 170	Medium to large, 100 volt, GPEAC with high-speed iterative operation and hybrid interface.	$50,000–$500,000
	1964	Ci 5000	Large fully transistorised, 10 volt GPEAC, with digital facilities and hybrid interface.	$200,000–$600,000
	1965	Ci 550	Large fully transistorised 100 volt GPEAC, with digital facilities and hybrid interface.	$40,000–$225,000
	1965	Ci 175	Medium capacity, 100 volt fully transistorised GPEAC with digital logic operations.	$40,000–$100,000
	1965	Ci 150	Medium capacity, low cost, fully transistorised, 100 volt GPEAC.	$—
Computer Systems Inc, Monmouth Junction, NJ	1961	DYSTAC 5800	Large high-speed iterative operation GPEAC, expandable to 134 op-amp system with dynamic storage capability	$20,000–$100,000
	1964	DYSTAC SS-100	Large fully transistorised high-speed iterative operation GPEAC, with digital facilities	$—
DIAN Laboratories Inc New York, NY	1959	DIAN 60	Range of 100 volt GPEAC systems, with 60-180 op-amps.	$—

TABLE 5.11 ELECTRONIC ANALOGUE COMPUTERS, MANUFACTURERS AND MODELS, 1960 TO 1970[40] (CONT'D)

Manufacturer	Date Introduced	Computer Models	Description	Price
Electronic Associates Incorporated, Long Branch, NJ	1958	DIAN 120 DIAN 180 PACE 231 R	Range of medium to large GPEACs, single-shot, or high-speed repetitive operation, and ADIOS option.	$20,000–$1,000,000
	1959	PACE TR-10	Small desk-top fully transistorised GPEAC, with 10 volt operation, solves 10th order differential equations	$4,000–$11,000
	1963	PACE TR-20	Small desk-top fully transistorised 10 volt GPEAC, 24 op-amps.	$4,500–$12,000
	1961	PACE TR-48	Desk-top fully transistorised 10 volt GPEAC, 48 op-amps.	$8,000–$35,000
	1962	HYDAC 2000	PACE 231 R, 100 volt GPEAC combined with EAI 350 series digital computer console for hybrid operation	$75,000–$650,000
	1963	HYDAC 2400	PACE 231 R, 100 volt GPEAC combined with EAI/Computer Control Co DDP-24 digital computer console for hybrid operation.	$170,000–$1,000,000
	1965	EAI 8800	Large fully transistorised GPEAC (replacement for valve 231 R series) with digital facilities and integral hybrid interface, 100-volt operation	$95,000–$495,000
	1965	EAI 8900 Hybrid Computer System	EAI 8800 combined with EAI 8400 digital computer via EAI 8930 Linkage, fully transistorised computer with 100-volt operation.	$700,000–$1,300,000

Manufacturer	Date Introduced	Computer Models	Description	Price
	1965	EAI 680	Medium size fully transistorised GPEAC with digital facilities and integral hybrid interface, 10-volt operation	$35,000– $200,000
Electronic Associates Incorporated, Long Branch, NJ	1966	EAI 690 Hybrid Computer System	EAI 680 combined with EAI 640 digital computer via EAI 639 Linkage, fully transistorised computer with 10-volt operation.	$110,000– $500,000
	1967	EAI 580	Medium size fully transistorised GPEAC with digital facilities, 10-volt operation	$10,000– $69,000
	1967	EAI 590 Hybrid Computer System	EAI 580 combined with EAI 640 digital computer via EAI 639 Linkage, fully transistorised computer with 10-volt operation.	$90,000– $350,000
	1968	EAI 7800	Large fully transistorised GPEAC with digital facilities and integral hybrid interface, 100-volt operation	$50,000– $300,000
	1968	EAI 7900 Hybrid Computer System	EAI 7800 combined with EAI 8400 digital computer via EAI 8930 Linkage, fully transistorised computer with 100-volt operation.	$500,000– $1,100,000
	1968	EAI 380	Small fully transistorised computer with digital facilities, 10-volt operation.	$8,000– $29,000
Geo Space Corp, Houston, Tex. (acquired Computer Systems Inc)	1967	DYSTAC SS-100	Large fully transistorised, analog/hybrid computer, 100 volt	$50,000– $350,000
GPS Instrument Co, Newton, Ma	1958	GPS Statistical Analogue Computer	Expandable repetitive operation GPEAC, with statistical data analysis capability, operates on a 3000:1 compressed time scale.	$20,000– $200,000

TABLE 5.11 ELECTRONIC ANALOGUE COMPUTERS, MANUFACTURERS AND MODELS, 1960 TO 1970[40] (CONT'D)

Manufacturer	Date Introduced	Computer Models	Description	Price
	1961	GPS Iterative Analogue Computer	High-speed, compressed time scale, iterative GPEAC	$35,000–$350,000
	1965	GPS 10000	Large high-speed GPEAC with iterative, statistical and digital facilities	$—
	1965–1969	GPS 240	Medium size GPEAC with digital facilities and 10-volt operation	$24,990–$49,000
		GPS 290T	High-speed GPEAC with DEC PDP-8 digital computer.	$95,000–$200,000
		GPS 390T	High-speed GPEAC with digital facilities.	$25,000–$200,000
		GPS 10,000T	Large fully transistorised, analogue/hybrid computer, 100-volt operation.	$50,000–$200,000
Hybrid Systems Inc, Anaheim, Calif	1968	SS-50	Medium size, transistorised, 100 volt, GPEAC with digital facilities and hybrid interface	$—
Milgo Electronic Corp, Miami, Fl.	1963	Model 4100 Series 7000	Large fully transistorised computer with digital facilities and integral hybrid interface, 100-volt operation.	$—
George A. Philbrick Researches Inc, Boston, Mass.	1958	K5, K7 Series	Plug-in repetitive-operation electronic analogue computing devices; expandable to form computers of various sizes.	$350–$220,000
Systron-Donner Corp, Donner Division, Concord, Calif.	1960–1967	SK2, SK5 Series		
	1960	Donner 3500	Small, 5 op-amp, lap-top GPEAC	$1,200–$1,800

TABLE 5.11 Electronic Analogue Computers, Manufacturers and Models, 1960 to 1970[40] (CONT'D)

Manufacturer	Date Introduced	Computer Models	Description	Price
	1961	Donner 3100	Small GPEAC solves 15th order differential equations	$12,000–$20,000
	1961	Donner 3200	Small to large, 100 volt GPEAC system	$5,000–$100,000
	1962	SD 20	Desk top, fully transistorised, 100 volt GPEAC	$8,000–$15,000
	1964	SD 40	Medium size, fully transistorised, 100 volt GPEAC	$15,000–$30,000
	1964	SD 80	Medium size, fully transistorised, 100 volt GPEAC	$20,000–$50,000
	1967	SD 80H	Fully transistorised GPEAC/hybrid with digital interface and 100-volt operation	$17,000–$90,000
Reeves Instrument Corp, New York, NY	1955	REAC 400	20 op-amp expandable GPEAC with optional digitally controlled input/output system	$20,000–$1,000,000
	1963	REAC 500	40 op-amp expandable GPEAC with digital facilities.	$30,000 up
TRW Systems, Redondo, Beach, Ca	1969	CDC 3100/ COMCOR CI-5000	Hybrid computer system combining COMCOR CI-5000 GPEAC with CDC 3100 digital computer.	$—
		DC 3100/ Beckman 2132	Hybrid computer combining Beckman 2132 GPEAC with CDC 3100 digital computer.	$—

consisted of a fully integrated analogue and small general-purpose digital computer, but was also designed to be used in conjunction with separate scientific digital computers, such as EAI's 8400, for large-scale hybrid computation.

From the mid-1960s there was a steady stream of new hybrid-compatible analogue computers from, among others, EAI, Comcor, Beckman, Systron-Donner and Applied Dynamics. These systems were themselves stand-alone hybrid computers, but they could also be linked to a host of different scientific digital computers to form large true-hybrid systems. The development of commercial stand-alone and true-hybrid computers was facilitated by the development of small scientific digital computers and medium-size and large digital computer systems with on-line capabilities. Large digital computers such as the CDC 6400, medium size systems such as the SDS 9300 and IBM 360/44, and small ones such as the SDS 930, Sigma 2

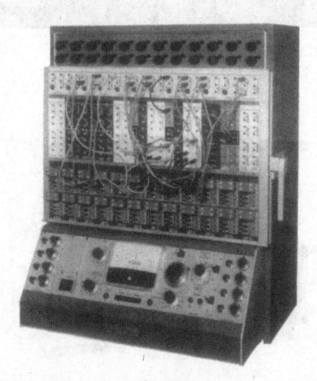

Figure 5.7   Electronic Associates Inc., EAI TR-20

FIGURE 5.8   ELECTRONIC ASSOCIATES INC., EAI TR-48

FIGURE 5.9   ELECTRONIC ASSOCIATES INC., HYDAC 2000

FIGURE 5.10   ELECTRONIC ASSOCIATES INC., EAI 690

FIGURE 5.11   ELECTRONIC ASSOCIATES INC., EAI 8945

Figure 5.12 Comcor Inc., Ci-175

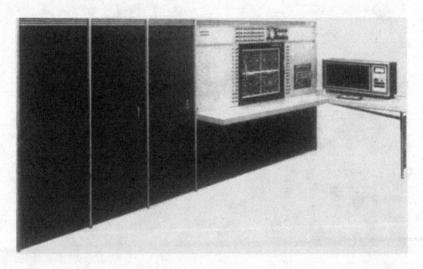

Figure 5.13 Comcor Inc., Ci-5000

FIGURE 5.14   COMPUTER SYSTEMS INC., SS-100

and the Digital Equipment Corporation's PDP 8, were used in hybrid computer systems.

Though EAI developed its own range of scientific digital computers, it was more generally the case that analogue/hybrid computer firms would either choose to team up with an existing digital computer manufacture to offer a complete hybrid system, or leave the choice of the digital computer to the user. Beckman, for example, chose not to develop its own range of scientific digital computers, but instead formed an alliance with Scientific Data Systems and developed a range of Beckman/SDS hybrid computers. More typically, the Comcor Ci 5000 and Applied Dynamics' AD 4 and AD 5 hybrid computers were designed so that they could be interfaced to almost any commercial scientific digital computer. For a list of

FIGURE 5.15 APPLIED DYNAMICS INC., AD 4

commercial electronic analogue and hybrid systems from 1960 to 1970, see Table(s) 5.11.

THE INFLUENCE OF NASA AND SPACE EXPLORATION ON THE ELECTRONIC ANALOGUE AND HYBRID COMPUTER INDUSTRY

The first successful launch of an earth satellite by the Soviet Union in 1957, and the subsequent launch of the first manned space rocket in 1959, led to the rapid acceleration of the American space exploration programme. In 1958, NACA was superseded by the civilian National Aeronautics and Space Administration (NASA), and expenditure on space exploration increased from several million dollars in 1958 to nearly $6 billion in 1963.[41] The massive US space programme provided the analogue computer industry with its next major fillip. In the late 1950s and 1960s NASA purchased large electronic analogue and hybrid computer systems for its research laboratories, as did the firms that NASA contracted to perform research and development.

These computers were used in many diverse applications, including the design of rocket engines, the simulation of docking manoeuvres and the simulation of moon vehicles for the Apollo programme.

In a sense, where the problems of cost, complexity and hazard in ICBM development and simulation left off, the manned space programme took over. However, in the space programme, total system simulation had an additional problem: that of including an astronaut in the control loop. Along with the enormous cost of rocket launches, the high profile and prestige of the space programme meant that rockets would not be test-flown with astronauts on board, either to train them or to iron out aerodynamic or control problems. Simulation for man–machine interaction had to be in real time and include as much of the actual system as possible. Thus, during the 1960s technical, economic and safety imperatives combined to stimulate, and sustain the development of commercial hybrid computer systems.[42]

EAI, Beckman Instruments Inc., GPS Instrument Inc., as well as new firms such as Comcor/Astrodata Inc., benefited greatly from the increase in demand for computer equipment that the race for space generated. EAI, already well established as a supplier of equipment to government-funded research facilities, benefited in particular from the sharp rise in demand for new equipment for NASA's research laboratories. In June 1960 EAI won a $1.51 million contract to supply NASA's Langley Research Centre with five fully expanded PACE 231-R computers. These computers were to be used on aspects of the Mercury project and to simulate space station rendezvous, satellite and missile launches. In 1960, contracts awarded by NASA to EAI for analogue computer and data-plotting equipment amounted to $2.49 million, and EAI was ranked fifth in NASA's list of top ten space contractors.[43] In 1960 NASA also ordered EAI PACE 231-R systems for its Goodard Space Flight Centre, Washington DC, its Space Task Unit at the Langley Research Centre and its Ames Research Centre at Moffet Field, California, and in 1961 EAI delivered a PACE 231-R to NASA's new Manned Space Flight Centre at Houston in Texas.[44]

However, EAI was not the sole supplier of electronic analogue computer equipment to NASA. From the mid-1960s NASA began to replace the electronic analogue computer systems it bought in the late 1950s and early 1960s. It replaced these with commercial hybrid computers manufactured by Beckman/Scientific Data Systems Inc., Comcor/Astrodata Inc. and Applied Dynamics as well as by EAI. For example, in 1964 NASA bought a Beckman/SDS hybrid computer system at a cost of $1 million for the Manned Spacecraft Centre at Houston. This large, all-transistor true-hybrid computer system was used in the Gemini and Apollo programmes for real-time simulation studies.[45] In 1967 NASA also bought one of EAI's latest and largest hybrid computer systems—the EAI 8900—for its Houston Manned Spacecraft Centre. The EAI 8900 was employed in investigations of the critical docking manoeuvres between the Apollo lunar landing module and the orbiting command ship. In the mid-1960 NASA bought an EAI 8800 hybrid computer for its Ames Research Laboratory at Moffet Field for simulation studies associated with the Apollo programme and the Saturn V rocket system. In the mid-1960s NASA also hired hybrid computer equipment at EAI's Computation Centre in Washington DC, to analyze the re-entry trajectories for the Apollo 8 mission.[46]

From the late 1950s to the mid-1970s private firms working on the space programme also invested in new electronic analogue and hybrid computer systems. For example, in 1962 the McDonnell Aircraft Corporation bought EAI's first HYDAC 2000 (Hybrid Digital/Analogue Computer). At McDonnell it was used in the development of control systems for NASA's Gemini project.[47] In the early 1960s, United Aircraft installed a large Beckman EASE analogue computer, which they used to simulate the Pratt & Whitney RL-10 rocket engine. The RL-10 was being developed by United Aircraft for NASA to power the upper stages of the Centaur and Saturn space vehicles.[48] In the early 1960s Boeing bought a Beckman EASE for its Aerospace Division in Huntsville, Alabama, to perform hybrid simulation studies of the launch phase of large space vehicles. And in 1967 the Lockheed Missile and Space Company installed a vast hybrid computer system, consisting of four Comcor Ci 5000 analogue/hybrid computer consoles and a CDC 6400 digital computer, at its plant in Sunnyvale, California. The system cost more than $4 million and was intended primarily for research related to NASA's Apollo space programme.[49] Indeed, by the end of the 1960s almost every major aerospace firm working on the space programme had purchased at least one of the new commercial hybrid computer systems which had started coming onto the market in the mid-1960s.

DEVELOPMENTS IN THE ELECTRONIC ANALOGUE AND HYBRID COMPUTER INDUSTRY 1965–1975: COMPETING WITH ELECTRONIC DIGITAL COMPUTERS

In the mid-to-late 1960s there was a great flurry of activity in the field of hybrid computation and its use grew rapidly, as did sales. By the end of the decade there were more than ten firms manufacturing hybrid computer systems. Yet the flurry of activity was also accompanied by increased competition from scientific digital computers.

All aspects of digital computing had undergone considerable improvement since the late 1950s. Improvements included higher computing speeds, larger, faster memories, and the ability to use the computer on-line instead of through the closed-shop batch-processing method. Significant improvements in the operation of digital differential analyzers (DDA) had also been achieved. Packard-Bell Inc. (later the Raytheon Computer Co.) had introduced a high-performance DDA known as TRICE (Transistorized Real-time Incremental Computer Expandable), in the early 1960s.[50] Significant improvements in software also made digital computers more appealing. FORTRAN was almost ten years old and was widely accepted and used in engineering applications, and improvements in many aspects of digital computer languages and programming methods made digital computers more "user-friendly". There had also been improvements in numerical analysis techniques and integration algorithms which reduced computational errors, and floating-point arithmetic obviated the need to "scale" equations. The big digital computer firms, such as IBM, undertook research to develop and improve specialised all-digital simulation languages for non-time-critical applications. Two of the most widely used were CSSL (Continuous System Simulation Language) and IBM's CSMP (Continuous Simulation Modelling Programme). These languages were similar to FORTRAN, but were developed to make the digital computer resemble an analogue computer from the point of view of the programmer.[51]

Improvements in digital computing meant that for many applications they were by far the most appropriate choice, even compared with hybrid computer systems. As more and more users chose digital computers over analogue ones, the market for analogue computers contracted. In the late 1960s, reductions in Federal and industrial funding for space and military R&D placed greater strain on the relatively small and highly competitive analogue computer market.[52] Yet while sales of the newer hybrid computer equipment continued to rise, sales of analogue computers that were not fully hybrid-compatible decreased. For this reason, certain parts of the analogue/hybrid computer industry faced rather more financial pressure than others. In the late 1960s, several of the long-running analogue computer manufacturing firms had little choice but to discontinue electronic analogue/hybrid computer manufacture. In 1967 George A. Philbrick Researches Inc. was taken over by the California-based electronics company Teledyne Inc., which soon phased out Philbrick's analogue computer products. Financial problems also forced Reeves, Beckman and the GPS Instrument Co. out of business.

Yet hybrid computing, as well as several of the specialised analogue/hybrid computer manufacturing firms, not only survived but for a time prospered, and there were even a few new firms to add to the list of manufacturers. There were a number of reasons for the persistence and success of hybrid computers some technical, some economic and some political.

The main advantage that hybrid computers offered over all-digital computers was analogue computing speed.[53] Real-time computation was essential in simulation studies where hardware, be it from a guided missile, aircraft or Apollo command module, was included in the control loop. For some applications, digital computers were fast enough for real-time hardware-in-the-loop simulation, but in large and complex dynamic problems there was generally insufficient computing capacity left over to perform the computation for other aspects of the simulation in real time.

Hybrid speed was also advantageous in problems with random variables where a large number of solution were required for statistical analysis. Yet many of these statistical studies could be performed equally well on a digital computer. In applications of this kind, it was largely economic factors that provided the justification for using hybrid computing. The all-digital solution of large, complex problems was generally much slower, and this led to higher overall costs per solution. For a problem requiring many thousands of solutions to be calculated, the high-speed hybrid computers enabled considerable cost savings.[54]

Notwithstanding these benefits of hybrid computing, one of the most important factors in the growth and maturation of commercial hybrid computers was the US government's endorsement of hybrid computation/simulation in research and development associated with the Apollo space programme. In the mid-1960s the US government played a direct role in promoting the use of hybrid computers by insisting that its contractors used real-time simulation. To this end, the government provided financial support for firms either to develop, purchase or use hybrid computers in their government-funded R&D.[55] NASA and the major aerospace firms bought the new hybrid systems in significant numbers, and by the late 1960s every major aerospace firm working on the space programme was using hybrid computation in a vast range of applications, including the simulation of space vehicle docking, engine and navigation system design, and astronaut training.

In the late 1960s analogue/hybrid computer sales in the USA peaked at approximately $50 million per annum. Yet the industry's reliance on government-funded military R&D and on the aerospace industry became apparent when budget reductions at NASA and the Pentagon led per annum sales of analogue/hybrid equipment to drop sharply to approximately $25 million in 1973, and to remain at between $25 to $30 million through to 1975.[56] Another significant factor in the sharp decline in hybrid computer sales was the increasing success of digital computer manufacturers at winning orders for their equipment in application environments, where hybrid computers were also a viable option. In applications such as hardware-in-the-loop simulations, iterative techniques for solving large systems of partial differential equations, statistical optimisation of dynamic systems, and Monte Carlo methods applied to nonlinear dynamic systems, the high speeds offered by hybrid computers gave them a considerable advantage over all-digital systems. Yet in applications where the hybrid's advantage was primarily economic, economics became a secondary consideration compared with the transparency and mathematical flexibility of the digital computer.

HIGH EXPECTATIONS, FALSE HOPES AND DECLINE: THE 1970s AND BEYOND

It was estimated that by 1970 at least 75% of simulation was being performed on all-digital systems.[57] This was the case even though studies had shown that hybrid computing offered speed advantages over the fastest digital computer of approximately 100:1, and also that considerable cost reductions could be achieved through the use of hybrid computers.[58] By the mid-1970s even the most ardent proponents of analogue computers generally agreed that the future for analogue computers lay either in the form of small desktop devices for education and/or as a fully subservient/integrated part of a digital-computer-controlled hybrid computer system.[59]

It was the potential of the latter of these two options to bring substantial economic benefits that prompted the Army Material Command (AMC) in 1972 to instigate a review of hybrid computer systems. This rejuvenated interest in the future of hybrid computing. The AMC's review gave indications of substantial potential cost savings within AMC projects of 30:1 compared with all-digital solutions. This encouraged the AMC to continue its investigation and to draw up proposals for what it referred to as Advanced (or fourth-generation) Hybrid Computer Systems (AHCS), including plans to integrate such systems into its computer network.[60] In 1973 the AMC commissioned further research to determine the relative speed and cost comparisons of digital and hybrid computing. Again, these studies indicated the considerable speed and economic benefits of employing AHCS.[61]

To gauge the overall potential for AHCS in the USA, the AMC conducted a survey of industry and academia in 1975. The survey indicated that the decline in hybrid computer sales had "bottomed out" and that there was significant demand for Advanced Hybrid Computer Systems, amounting to a potential market of $300 million in the USA alone. The survey's conclusion was optimistic:

> The Advanced Hybrid Computer System presents a viable alternative to high-priced systems of digital computers in solving dynamic problems in the scientific community of tomorrow. The AHCS will solve problems faster, cheaper and in many instances more optimally than even the next generation of digital computers, thus giving scientists and engineers the efficient and effective tool they need to perform their mission in the best possible way.[62]

However, though the AMC generated some increased interest in hybrid computing and stimulated the commercial manufacturers to re-address the future of their field, the anticipated financial support of research and development did not materialise, nor did the anticipated rejuvenation of the hybrid computer market.[63] By the late 1970s the apogee for commercial hybrid computers had been passed, and the industry swept towards digital products in order to remain in business. EAI, which for almost twenty years had been the market leader in analogue and subsequently hybrid computers, was one of the few firms to survive beyond the mid-1970s. In the late 1970s EAI continued to manufacture a range of small-analogue and large hybrid computer products, including the PACER 2000 (Precision Analogue Computer Equipment, Revised): a hybrid computer which incorporated analogue microchip technology.[64]

Unlike its major competitors, EAI kept analogue computing elements and techniques within its products alongside digital parallel-processing components and methods. In the early 1990s EAI was still designing and manufacturing computers, as was Applied Dynamics Inc. Applied Dynamics was one of the firms that chose the all-digital route and in 1977 launched the AD 10 digital computer, marking its departure from analogue computer development. In the early 1990s, both EAI and Applied Dynamics Inc. were still in business.

To a limited extent, electronic analogue computation survives, as small systems for use in education, and embedded in hybrid computers, few of which outwardly reveal any non-digital computer characteristics. In 1989 it was estimated that there were still approximately three hundred hybrid computer installations in operation around the world.[65]

CONCLUSION

The underlying trend in the American electronic analogue/hybrid computer industry was away from in-house development to the purchase of systems from the increasingly specialised manufacturing firms. In the USA during the 1950s and 1960s the manufacturing and user base grew, and electronic analogue and hybrid computers were used in an increasingly diverse range of applications: from economics, telecommunications, chemical engineering and textiles, to medicine and the simulation of biological systems. However, the growth of a distinct postwar electronic analogue/hybrid computer industry cannot be explained in terms of the demand for aids to computations in these fields of study. The crucial and formative relationship was that between the computational characteristics and high speed of the electronic analogue computing, and the economic and technical imperatives associated with aeronautical design and the development of aircraft, missiles and space vehicles. This relationship was undermined by extraordinary improvements in all aspects of digital computing performance, combined with substantial reductions in the cost of digital computing equipment and their widespread adoption.

NOTES

1. This classification scheme is also used in chapter 6 for the British data.

2. The Sperry Gyroscope Company, of Great Neck NJ, main interests lay in the manufacture of navigation systems and ordnance fire-control equipment. In the immediate post-war period Sperry developed their own electronic analogue computer. Yet, Sperry did not make the transition from developer/user to manufacturer. Sperry did further develop and build copies of its own analogue computer system. However, after commercial systems became available Sperry, like many other developer/users, bought electronic analogue computer systems rather than developing new equipment of their own. By 1952, in addition to their own electronic analogue computer system, Sperry was using equipment manufactured by Reeves and GAP/R Inc. There were two Reeves REAC computer installations at Sperry. Both of which were located in Sperry's Aeronautical Engineering Division. One of the REACs was owned by Sperry, the other was on loan from the US Navy for use by the Sparrow missile development group at Sperry. Sperry's Aeronautical Engineering Division also owned a small (8 op-amp) repetitive operation analogue computer manufactured by GAP/R Inc. In the Spring of 1953, Sperry's Surface Armament Division bought a Goodyear GEDA computer system. By 1955 six REAC computers were in use at Sperry Gyroscope. G. A. Korn and T. M. Korn, *Electronic Analogue Computers (D-C Analog Computers)*, McGraw-Hill, New York, 1952, pp. 336–337; P. M. Sherman, "A Report on DC Electronic Analogue Computation and the Aeronautical Division Analogue Computer Facility", *Internal report 5234–3522*, Sperry Gyroscope Co., Great Neck, NJ, 25 Aug 1955, pp. 1–148, Hagley Museum and Library, Sperry Collection; P. M. Sherman, "The Theory, Operation and Use of Philbrick High-Speed Analogue Computer", *Internal report 5234–3458*, Sperry Gyroscope Co., Great Neck, 2 Sept 1955, pp. 1–56, Hagley Museum and Library, Sperry Collection; Q. J. Hoffman, H. B. Sabin and H. J. Smith, "Requirements for Computing Facilities in the Sperry Engineering Division", *Internal report 5281–7142*, Sperry Gyroscope Co., Great Neck, 15 Jan 1953, pp. 1–26, Hagley Museum and Library, Sperry Collection.

3. J. M. Carroll, "Analog Computers for the Engineer", *Electronics*, June 1956, pp. 122–129, p. 122.

4. H. M. Paynter, "In Memorial: George A. Philbrick (1913–1974)", *Trans., of the ASME, Journal of Dynamic Systems, Measurement and Control*, June 1975, pp. 213–215; D. H. Sheingold "George A. Philbrick: Gentleman, innovator," *Electronic Design*, Vol. 3, 1 Feb 1975, p. 7.

5. Production of the K3 series began in 1948 and continued until 1958, when they were largely superseded by the K2 and K5 series. On the front of K3 unit, one of four input jacks

and two (one providing the inverse of the other) output jacks permitted computing signal connections. On the back, five pin input and output connectors supplied power connections and permitted power cable connections to be made in cascade between units. A calibrated dial on the front of each unit enabled the magnitude of characteristics to be set, and a lamp was used to indicate when an overload condition occurred. The input and output signal range was +/–50 volts. A = adder, B = limiter, C = multiplyer, H = backlash, J = integrator, L = lag, Z = delay. *The Lightning Empiricist*, George A. Philbrick Researches Inc, Boston, Mass., Vol. 1, no. 1, June 1952.

6.   In 1950 Philbrick entered into partnership with Henry M. Paynter and together they formed the Pi-Squared Engineering Company and Established the American Centre for Analog Computing, on Clarendon Street, Boston. The centre provided a consultancy service to industry and academia, and acted in part as a showcase for GAP/R analogue computing products. Customers could come to the centre either to run problems on the computer facility located there, or seek advice on the theory and practice of how to use analogue computational methods to solve their problems in engineering and science. Personal Interview: Henry M. Paynter, Pittsford, Va, USA, 27 Oct 1990.

7.   H. M. Paynter (ed.), *A Palimpsest on the Electronic Art*, George A. Philbrick Researches Inc, Boston, Mass., 1955; G. A. Philbrick and C. E. Mason, "Automatic Control in the Presence of Process Lags," *Trans.*, ASME, Vol. 62, May 1940, pp 295–308; G. A. Philbrick, "Designing Industrial Controllers by Analog," *Electronics*, Vol. 21, June 1948, pp. 108–111; G. A. Philbrick and H. M. Paynter, "The Electronic Analog Computor as a Lab Tool," *Industrial Laboratories*, Vol. 3, no. 5, May 1952, pp. 32–37.

8.   D. M. MacKay and M. E. Fisher, *Analogue Computing at Ultra-high Speeds*, John Wiley and Sons, New York, 1962; A. E. Rogers and T. W. Connelly, *Analog Computation in Engineering Design*, McGraw-Hill, New York, 1960; R. Tomovic and W. J. Karplus, *High-speed Analog Computers*, John Wiley and Sons, New York, 1962.

9.   P. A. Holst, "Sam Giser and the GPS Instrumentation Company: Pioneering Compressed-Time Scale (high-speed) Analog Computing", in J. McLeod (ed.), *Pioneers and Peers*, The Society for Computer Simulation International, San Diego, Ca., 1988, pp. 4–10.

10.   NASA was one of EAI's biggest customers. For example in June 1960 NASA awarded EAI a contract for $1.51 million. This was the single largest contract received by EAI in the fifteen years since the company was founded. See section 5.8 below. *EAI Annual Report 1960*, Electronic Associates Inc, Long Branch, NJ, 1960.

11.   The prototype systems developed at UCLA had 30 amplifiers which could be adapted to perform various mathematical functions in real time. *Special Products News: The Beckman EASE Computer*, Beckman Instruments Inc., South Pasadena, Ca., 18 April, 1952; *EASE Computer*, Berkeley Division, Beckman Instruments Inc., Richmond, Ca., c. 1954; Anon., *Electronic Analogue Computation Laboratory: Bulletin No. 2*, Department of Engineering, UCLA, Ca., 12 Dec 1950.

12.   *Moody's Industrial Manual*, New York, 1954, p. 245.

13.   This was soon superseded by the EASE 1000 Series, costing between $6,000 and $10,000, and subsequently by the EASE 1100 Series. *Special Products News: The Beckman EASE Computer*.

14.   J. M. Carroll, "Analog Computers for the Engineer".

15.   Anon., *The Simulation Council Newsletter*, in *Instruments & Control Systems*, Vol. 32, May 1959, p. 742.

16.   Anon., *The Simulation Council Newsletter*, in *Instruments & Control Systems*, Vol. 32, July 1959, p. 1057; *Dian Laboratories Inc*, Catalogue, Dian Laboratories Inc, New York, 1956.

17.   *Donner Model 30 Analogue Computer*, Catalogue, Donner Scientific Co., 1955; *Donner Model 3000 Analogue Computer*, Catalogue, Donner Scientific Co., 1957.

18. L. B. Wadel and A. W. Wortham, "A Survey of Electronic Analog Computer Installations", *Trans., IRE, Electronic Computers*, Vol. EC-4, June 1955, pp. 52–55.

19. For an analysis of the rapid growth of aerospace and electronics in this region see, A. J. Scott, "The aerospace-electronics industrial complex of Southern California: The formative years, 1940–1960", *Research Policy*, Vol. 20, 1991, pp. 439–456.

20. L. B. Wadel and A. W. Wortham, "A Survey of Electronic Analog Computer Installations", p. 54.

21. Ibid., p. 54.

22. In nuclear engineering, analogue computers were used in reactor design, for the solution of equations associated with the dynamic behaviour of power levels, reactor heat transfer, coolant flow and steam-systems kinetics. Analogue computers have also been used for the complete simulation of nuclear power plants. Anon., "A Nucleonics Survey: Analog Computer Use in the Nuclear Field", *Nucleonics*, Vol. 15, no. 5, May 1957, p. 88; S. O. Johnson, and J. N. Grace. "Analog Computation in Nuclear Engineering", *Nucleonics*, Vol. 15, no. 5, May 1957, pp. 72–75, p. 73.

23. Anon., *Instruments & Automation*, Vol. 30, Nov 1957, p. 2089.

24. Anon., *Instruments & Control Systems*, Vol. 34, Aug 1961, p. 1478.

25. The start of production of the REAC 400 at the Reeves New York City and Mineola NY plants began in 1955. The first delivery went to Farnsworth Electronic Co. Fort Wayne, Ind. Farnworth had recently recieved $10M worth of orders for control and test equipment for the goverment guided missile programme. *The Wall Street Journal*, New York, Thursday, 28 July 1955. p. 3.

26. Lockheed's Georgia Division's Mathematical Analysis Dept., had a 130 amplifier analogue computer system, and the Flight Controls Research Laboratory a 30 amplifier unit. Anon., *Instruments & Control Systems*, Vol. 34, April 1961, p. 686; L. K. Yoskowitz, "Methods of Pre-flight and Post-flight Simulation" *The Simulation Council Newsletter*, in *Instruments & Control Systems*, Vol. 34, April 1961, pp. 681–682; W. C. Bennett "The Interconnection of Remotely Located Analog Computer Equipment", *The Simulation Council Newsletter*, in *Instruments & Control Systems*, Vol. 34, Aug 1961, p. 1477.

27. Anon., "EAI Computation Centre in Los Angeles Serves West", *Western Electronic News*, December 1956, p. 1–4, p. 3. EAI opened its new Los Angeles Computation Centre in the heart of the aircraft industry at 1500 East Imperial Highway, El Segundo. It was located in a new purpose-built building and Lloyd F. Christianson, the President of EAI, stated that "the establishment of the LA Computation Centre is an endeavour to bring the benefits of analog computation within the financial reach of all industry in the west coast area, regardless of size, on an hourly rental basis." Ibid., p. 3.

28. Where problems arose using conventional analogue or digital techniques, ways had been devised to minimise these. For example, one common practice in the early 1950s was to use separate analogue and digital computers to provide check solutions. It was also the case that for large, complex or problematic designs, separate analogue and digital computers were used to study different parts of the same design problem.

29. By 1956 the analogue computer laboratory at Convair San Diego contained 728 amplifiers. It was organized into 17 separate analogue computer consoles located in 3 rooms on two floors. Anon., *Simulation Council News Letter*, The Simulation Council, Camarillo, Ca., May 1953, pp. 4–7; Anon., *Simulation Council News Letter*, in *Instruments & Automation*, Vol. 29, Aug 1956, p. 1567; Stanley Rogers, Typescript of Interview, 9 Aug 1973, Smithsonian Computer History Project—Oral History Collection, National Museum of American History Archive Centre, Washington DC; Anon., "Biographies of Significance: Stanley Rogers", *Instruments & Automation*, Vol. 28, Nov 1955, p. 1847.

30. E. M. Emme, *Aeronautics and Astronautics*, National Aeronautics and Space Administration, Washington DC, 1961; R. L. Perry, "The Atlas, Thor, Titan and

Minuteman", in E. M. Emme, (ed.), *The History of Rocket Technology*, Wayne State University Press, Detroit, 1964, pp. 142–161; W. von Braun and F. I. Ordway III, *History of Rocketry & Space Travel*, Nelson, London, 1967, pp. 132–134.

31.. The committee members were selected by Trevor Gardner, the Special Assistant for Research and Development to the Secretary of the Air Force. Already convinced by staff from Convair of the feasibility of ICBMs, Gardner chose prominent figures, like John von Neumann, who were sympathetic to his own view and would not only endorse a pro- gramme of accelerated ICBM development, but would lend credence to the decision and thus generate support for it. See, B. D. Adams, *Ballistic Missile Defence*, American Elsevier, New York, 1971; W. von Braun and F. I. Ordway III, *History of Rocketry & Space Travel*, pp. 120–149; D. Mackenzie, *Inventing Accuracy: A Historical Sociology of Nuclear Missile Guidance*, MIT Press Cambridge, Mass., pp. 103–113; E. G. Schwiebert, *A History of the US Air Force Ballistic Missiles*, Frederick A. Praeger, New York, 1965; H. York, *Race to Oblivion: A Participant's View of the Arms Race*, Simon and Schuster, New York, 1970.

32. Anon., *Simulation Council News Letter*, in *Instruments & Automation*, Vol. 29, Aug 1956, p. 1567; J. L. Greenstein, "Application of ADDAVerter System in Combined Analogue–Digital Computer Operations", Paper Presented at *AIEE Pacific General Meeting*, Paper No. 56-842, June 1956; R. M. Leger, "Specification for Analogue–Digital–Analogue Converting Equipment for Simulation Use", Paper Presented at *AIEE Pacific General Meeting*, Paper No. 56-860, June 1956; R. M. Leger, "Requirements for Simulation of Complex Systems", *Proc., Flight Simulation Symposium*, Sept 1957, pp. 125–131; J. H. McLeod and R. M. Leger, "Combined Analog and Digital Systems—Why, When and How", *Instruments & Automation*, Vol. 30, June 1957, pp. 1126–1130; A. N. Wilson, "Recent Experiments in Missile Flight Dynamics Simulation with the Convair 'ADDAVerter' System," *Proc., Combined Analogue–Digital Computer Systems Symposium*, Philadelphia, Pa, Dec 1960; A. N. Wilson, "Use of Combined Analogue–Digital Systems for Re-entry Vehicle Flight Simulation", *Proc., Eastern Joint Computer Conference*, Washington DC, Dec 1961, pp. 105–113.

33. W. von Braun and F. I. Ordway III, *History of Rocketry & Space Travel*, pp. 132–134.

34. Anon., *Simulation Council News Letter*, in *Instruments & Automation*, Vol. 29, Feb 1956, pp. 301–304; W. F. Bauer and G. P. West, "A System for General-Purpose Analog–Digital Computation", *Journal of the Association for Computing Machinery*, Vol. 4, Jan 1957, pp. 12–17; W. F. Bauer, "Aspects of Real-Time Simulation", *Trans., IRE, Electronic Computers*, EC-7, 1958, pp. 134–136; G. A. Bekey and W. J. Karplus, *Hybrid Computation*, Wiley, New York, 1968, p. 154.

35. C. G. Blanyer and H. Mori, "Analogue, Digital and combined Analogue-Digital Computers for Real-Time Simulation, *Proc., Eastern Joint Computer Conference*, Washington DC, Dec 1957, pp. 104–110; F. A. Brown, "Hybrid Simulation Techniques", *Simulation Council News Letter*, in *Instruments & Control Systems*, Vol. 36, Sept 1963, pp. 145–146; A. J. Burns and R. E. Kopp, "Combined Analogue-Digital Computer Simulation", *Proc., Eastern Joint Computer Conference*, Washington DC, Dec 1961, pp. 114–123; M. Paskman and J. Heid, "The Combined Analogue–Digital Computer System", *Proc., Combined Analogue–Digital Computer Systems Symposium*", Philadelphia, Pa, Dec 1960; G. A. Paquette, "Progress of Hybrid Computation at United Aircraft Research Laboratories", *Proc., Fall Joint Computer Conference*, 1964, pp. 695–706; S. Shapio and L. Lapides, "A Combined Analogue–Digital Computer for Simulation of Chemical Processes", *Proc., Combined Analogue–Digital Computer Systems Symposium*", Philadelphia, Pa, Dec 1960; H. K. Skramstad, A. A. Ernst and J. P. Nigro, "An Analogue–Digital Simulator for the Design and Improvement of Man–Machine Systems, *Proc., Eastern Joint Computer Conf.*, Washington DC, Dec 1957, pp. 90–96; H. K. Skramstad, "Combined Analog–Digital Techniques in Simulation", in F. L. Alt and M. Rubinoff (eds.), *Advances in Computers*,

Academic Press, New York, Vol. 3, 1962, pp. 275–298; M. L. Stein, "A General-Purpose Analogue–Digital Computer System", *Proc., Combined Analogue–Digital Computer Systems Symposium*", Philadelphia, Pa, Dec 1960.

36. This development emerged from concerns among users of large electronic analogue computer systems that the proportion of the total solution time devoted to setting up and checking a problem was too high. Studies indicated that 70% of total computer time was spent on set up and checking. Consequently, in the mid-1950s the leading electronic analogue computer manufacturers began to develop digitally controlled input/output systems to automate these operations. These systems also made it easier to store the details of a problem setup, and thus reduced the time required to re-instate it at a later date. In the mid-1950s, the Reeves Instrument Co incorporated digitally controlled setup and checking equipment developed for the WADC computer into its standard range of REAC computer products. In 1956 Beckman and EAI also introduced systems of this kind. EAI gave their system the acronym ADIOS, for Automatic Digital Input-Output System. ADIOS included keyboard or paper-tape control of the setting of attenuators and potentiometers, could read back the current setting of pots and output voltages of all operating components, and read out the results of static and dynamic check procedures. The Beckman system, known as DO/IT (Digital Output/Input Translator), was very similar, and was incorporated into the EASE 1200 Series. H. E. Harris, "New Techniques for Analogue Computation," *Instruments & Automation*, Vol. 30, May 1957, pp. 894–899.

37. EAI's HYDAC 2400 launched the following year did however, contain a general-purpose digital computer: the Computer Corporation Inc's DDP-24.

38. Transistorized analogue computers were in themselves not new; EAI had in fact already introduced two all-transistor computers, the TR-10 in 1959 and the TR-48 in 1961. These computers were however, not designed for combined analogue–digital computation. *EAI Annual Report*, Electronic Associates Inc, Long Branch, NJ, 1961.

39. Ibid.

40. Sources: Anon., "Survey of Commercial Analog Computers", *Computers and Automation*, Vol. 10, June 1961, pp. 117–118; Anon., "Survey of Commercial Analog Computers", *Computers and Automation*, Vol. 11, June 1962, pp. 130–132; Anon., "Characteristics of General Purpose Analog Computers", *Computers and Automation*, Vol. 13, June, 1964, p. 78; Anon., "Characteristics of General Purpose Analog and Hybrid Computers", *Computers and Automation*, Vol. 18, 1969, pp. 151–183; Also *Instruments and Control Systems*, and *Simulation*, from 1960 to 1970, trade publications and annual reports.

41. C. D. Perkins, "Man and Military Space", *Journal of the Royal Aeronautical Society*, Vol. 67, no. 631, July 1963, pp. 397–412.

42. R. M. Barnett "NASA Ames Hybrid Computer Facilities and their Application to Problems in Aeronautics", *International Symposium on Analogue and Digital Techniques Applied to Aeronautics*, Liege, Belgium, Sept 1963; L. E. Fogarty and R. M. Howe, "Flight Simulation of Orbital and Reentry Vehicles", *Trans., IRE, Electronic Computers*, EC-11, Aug 1962, pp. 36–42; C. A. Jacobsen, "The Application of the Hybrid Computer in Flight Simulation", *Proc., IBM Scientific Computing Symposium on Computer-Aided Experimentation*, 1966, pp. 206–244; O. F. Thomas, "Analogue–Digital Computers in Simulation with Humans and Hardware", *Proc., Western Joint Computer Conf.*, 1961, pp. 639–644; T. D. Truitt, "Hybrid Computation ... What is it? ... Who Needs it? ...".

43. *EAI Annual Report*, Electronic Associates Inc., Long Branch, NJ, 1960, p. 3.

44. During 1960 more than 14 per cent of EAI's 231-R computer sales went to firms in the chemical and petroleum industries. Du Pont, Shell, Chemstrand, Monsanto, Humble, Dow Chemical. Ibid., pp. 7–8.

45. Anon., *Instruments & Control Systems*, Vol. 37, Nov 1964, p. 144.

46.   *EAI Annual Report*, Electronic Associates Inc., Long Branch, NJ, 1968; *Moody's Computer Industry Survey*, New York, Vol. 1, no. 1, 1965, p. 457, pp. 474–475.

47.   McDonnell also had two fully expanded 231R-V systems linked to a UNIVAC 1218 scientific digital computer and five EAI PACE 131R computers. J. L. Clancy and M. S. Fineberg, "Hybrid Computing: a users's view", *Simulation*, Aug 1965, pp. 104–112; *EAI Annual Report*, Electronic Associates Inc., Long Branch, NJ, 1962.

48.   The Pratt & Whitney RL-10 was the US's first liquid-hydrogen rocket engine. Work on it and on complex missile and aircraft stability problems, vibrational analysis and other design problems was undertaken by the Analogue Section at the United Aircraft Research Laboratories, East Hartford, Conneticut. Anon., "Research Leadership at United Aircraft and Beckman", *Instruments & Control Systems*, Vol. 35, 1961, p. 170.

49.   G. A. Bekey and W. J. Karplus, *Hybrid Computation*, p. 165. Another example of the use of hybrid simulation in the space programme was at the Space Division of the North American Rockwell Corporation. It was responsible for the design and manufacture of a significant proportion of the Apollo launch vehicle, including the command module, service module, launch escape system the Apollo space craft-lunar module adaptor and the Saturn II stage of the Saturn V rocket. In May 1966 the Space Division opened a new Flight Simulation Laboratory primarily to undertake evaluation studies of guidance and control systems for the Apollo command and service modules. The laboratory contained a full-scale Apollo command module mock-up as well as a flight table, engine simulator, and guidance and control system. There were two simulation computers, one digital and one analogue. The digital computer was a Scientific Data System 9300 and the analogue computer consisted of four interconnected EAI 231-Rs, containing more than 450 operational amplifiers. J. H. Argyris, "The Impact of Digital Computers on Engineering Science," *The Aeronautical Journal of the Royal Aeronautical Society*, Vol. 74, Feb 1970, pp. 111–127.

50.   Customers for TRICE included North America Aviation which installed a TRICE/250 in 1962, and in 1964 TRICE/440s were installed at NASA's Manned Spacecraft Centre, Houston, Texas, and Marshall Space Flight Centre, Huntsville Alabama. See, O. A. Reichardt, M. W. Hoyt and W. Thad Lee, "The Parallel Digital Differential Analyzer and its Applications as a Hybrid Computing Systems Element," *Simulation*, Feb 1965, pp. 104–126; G. A. Korn, "Progress of Analogue/Hybrid Computation", *Proc., Fifth International Analogue Computation Meeting*", Lausanne, 28 Aug–2 Sept 1967, pp. 22–34.

51.   D. Brennan and H. Sano, "Pactolus—a digital analog simulator program for the IBM 1620", *Proc., Fall Joint Computer Conference*, 1964, pp. 299–312; R. N. Linebarger and D. Brennan, "A Survey of Digital Simulation", *Simulation*, Vol. 3, Dec 1964, pp. 22–36; D. Tiechroew and J. F. Lubin, "Discussion of Computer Simulation Techniques and a Comparison of Languages", *Simulation*, Oct 1967, pp. 181–190.

52.   D. C. Mowery and N. Rosenberg, *Technology and the Pursuit of Economic Growth*, Cambridge University Press, Cambridge, 1989, p. 180

53.   G. A. Korn and T. M. Korn, *Electronic Analogue and Hybrid Computers*, 2ed., McGraw-Hill, New York, 1972, chapter 2.

54.   A. G. Edwards, *et al.*, *Advanced Hybrid Computer Systems*, US Army Materiel Command, Alexandria, Va, June 1973; V. J. Sorondo, R. B. Wavell and F. E. Nixon, "Cost and Speed Comparisons of Hybrid VS. Digital Computers", paper presented at the *Special Symposium on Advanced Hybrid Computing*, sponsored by US Army Materiel Command, at the *Summer Computer Simulation Conference*, San Francisco, Ca., 24 July 1975.

55.   Personal Interview: Dr. Walter J. Karplus, UCLA Computer Science Dept, Los Angeles, Ca., 9 Oct 1991. Dr. Karplus was an advisory to the US Government on scientific computing and simulation.

56.   *EAI Annual Report*, Electronic Associates Inc, Long Branch, NJ, 1970; Personal Interview: Derek J. Cartwright, Hendon, Sussex, 1 May 1991; D. C. Peak, "A Survey of Requirements

for a Present-Day and Fourth Generation Hybrid Computer Systems in the United States", paper presented at the *Special Symposium on Advanced Hybrid Computing*, sponsored by US Army Materiel Command, at the *Summer Computer Simulation Conference*, San Francisco, Ca., 24 July 1975.

57.   M. White, *et al.*, "Trends in Simulation Hardware and Application: A Panel Discussion", *Simulation*, June 1971, pp. 279–285.

58.   J. Clancy, "Notes on the 'Bandwidth' of Digital Simulation (Technical Comment)", *Simulation*, June 1967, pp. 19–20; A. G. Edwards, *et al.*, *Advanced Hybrid Computer Systems*, US Army Materiel Command, Alexandria, Va, June 1972.

59.   One of the advantages of the latter was that analogue/hybrid computer programming which was previously based on patch-board methods could be fully-automated this would reduce analogue computer costs and increase the overall reliability of the analogue/hybrid system.

60.   A. G. Edwards, *et al.*, *Advanced Hybrid Computer Systems*.

61.   L. Wolin, "The Wolin–Saucier–Peak (WSP) Scientific Mix: A Quantitative Method for Comparing Hybrid and Digital Computer Performance", paper presented at the *Special Symposium on Advanced Hybrid Computing*, sponsored by US Army Materiel Command, at the *Summer Computer Simulation Conference*, San Francisco, Ca., 24 July 1975.

62.   D. C. Peak, "A Survey of Requirements for a Present-Day and Fourth Generation Hybrid Computer Systems in the United States", p. 166.

63.   D. Miller, "Faster than real time", *Simulation*, Nov 1977, pp. 117–118.

64.   In the mid-to-late-1970s in the USA the four principal firms manufacturing electronic analogue/hybrid computers were Applied Dynamics Inc, Astrodata/Comcor Inc, Denelcor Inc, and Electronic Associates Inc.

65.   P. Landauer, quoted in L. Hassig (ed.) *Alternative Computers*, Time-Life Books, Alexandria, Va, 1989, p. 30.

CHAPTER 6

The Origins, Commercialisation and Decline of Electronic Analogue and Hybrid Computing in Britain, 1945–1975

INTRODUCTION

In the previous two chapters I described developments in electronic analogue computing in the USA. We have seen how the military played an important role in establishing and supporting a distinct postwar analogue computer industry. The overarching pattern was away from developer/users and user/manufacturers to commercial production by specialised analogue computer firms, and from electronic analogue computers to hybrid systems. In Britain a similar pattern can be seen, but with important differences in scale and in context. The smaller scale of the British electronic analogue computer industry can be explained largely in terms of the comparatively small scale of British programmes to develop guided missiles, military and civil aircraft, and space vehicles. Consequently, in Britain the role played by other industries, most notably the electrical and nuclear industries, was more significant than in the USA. In Britain we see that commercialisation of electronic analogue computers began in 1953, five years later than in the USA, and that there were fewer firms in this industry. Moreover, by the late 1950s British firms were facing stiff competition from the leading American firms, which began expanding into the European market.

THE ROYAL AIRCRAFT ESTABLISHMENT AND ELECTRONIC ANALOGUE COMPUTER DEVELOPMENT

Prior to the early 1950s, the bulk of the postwar R&D undertaken specifically to develop general-purpose electronic analogue computing systems was carried out at the Royal Aircraft Establishment (RAE), Farnborough. In the late 1940s and early 1950s the RAE was Britain's principal research establishment for guided weapons. It was funded by the Ministry of Supply which was responsible for the development and procurement of all new weapons for the British armed forces. Though some work on the development of guided missiles had taken place at the RAE during World War II, it was not until after the war that the RAE took up a central role. In postwar Britain, guided weapons were viewed as vital to future defence against air attack. However, in 1945 the organisation of research and development in guided projectiles and weapons was "highly differentiated and coordinated only at Cabinet level".[1] To improve its overall efficiency, the guided weapons programme was reorganised, and this led to the creation of the Guided Projectile Establishment (GPE) at Westcott, with Sir Alwyn Crow of the MoS as its Director. The GPE began operating in April 1946 and was responsible for all guided weapon research, except those guided weapons that would be launched from piloted aircraft. This task went to the RAE, Farnborough. Reorganisation within the RAE followed, and the various sub-divisions involved in missile design were brought together to form the Department of Controlled Weapons, with Dr George Gardner as the departmental head.[2] However, the division of responsibilities between GPE and RAE created many problems and much duplication of effort.[3] In 1947 the MoS once again reorganised Britain's guided weapons programme and the RAE was given principal responsibility for guided weapons R&D for the Army, Navy and Air Force. As a result, the Controlled Weapons Department at the RAE was reconstituted as the Guided Weapons Department.[4]

During the first postwar restructuring of the RAE which began in 1946, staff from the Ministry of Aircraft Production's Air Warfare Analysis Section joined the newly formed Controlled Weapons Department. One of those recruited was W. R. Thomas. Thomas later became head of the group responsible for electronic analogue computer development at the RAE. Thomas and his group drew upon analogue computing techniques developed at the government's Telecommunication Research Establishment during the war.[5] The first public indication that work on the development of electronic analogue computers was being undertaken at the RAE was given in 1948. That year, the RAE exhibited a small analogue computing system at the Physical Society's Thirty-Second Annual Exhibition of Scientific Instruments and Apparatus in London. The computer was designed by J. J. Gait and D. W. Allen of the RAE. It had only five DC operational amplifiers and was developed primarily as a proto-type for a larger computer system known as GEPUS (General Purpose Simulator).[6] Completed in 1950, GEPUS was the first large DC general-purpose electronic analogue computer system in Britain. It was designed jointly by M. Squires of the RAE and staff from an electronics firm—the Plessey Company—that was working on the project under a contract from the MoS.[7] Plessey were also contracted to manufacture and install GEPUS for the RAE. This was a one-off system and Plessey did not subsequently manufacture electronic ana-logue computers on a commercial basis.

The computational demands associated with missile design grew rapidly in the postwar years. Though in 1950 GEPUS was a rela-tively large system with 31 DC computing amplifiers, and in total contained 850 thermionic valves, it was not powerful enough to perform many of the computations and simulations required in the design and testing of guided missile systems. In particular, the simu-lation of a missile manoeuvring in three dimensions was a problem faced by designers that was beyond the capacity of GEPUS. As a result, a much larger computer facility was soon under construction at the RAE.

TRIDAC: The RAE's Three-Dimensional Analogue Computer

In January 1950 work on the development of a new guided missile simulation and design facility began at the RAE. This was called TRIDAC (Three-Dimensional Analogue Computer) and was designed jointly by staff from RAE, Farnborough and Elliott Brothers (London) Ltd, who were also responsible for its construction. TRIDAC took almost five years to complete at an estimated total cost to the Ministry of Supply of £750,000. At its launch in 1954 it was the largest com-puting facility of its kind in Britain. It occupied a total of 6,000 sq.ft.

of floor space, used 8,000 thermionic valves and had 650 DC operational amplifiers.[8] It was estimated that the computing capacity of TRIDAC was equivalent to 10,000 staff operating desk calculating machines.[9] Though this was an impressive statistic, the comparison with desk calculating machines tells us little about the real power of TRIDAC, which lay not only in its ability to solve large problems but to do so in real time. Computational speed was a crucial consideration in the choice of analogue rather than digital computing methods in the design of TRIDAC. The designers at Elliott and the MoS explained the reasoning behind this choice as follows:

> Digital machines can be programmed to carry out long series of operations in sequence, with great accuracy. However, with existing techniques, large numbers of operations cannot be carried out simultaneously (or sufficiently quickly when executed in sequence) so that in general an involved problem cannot be reproduced on a real time scale and thus the inclusion of parts of the real system in place of the machine calculation is not possible. It may be that in future digital machines will operate more rapidly so that the construction of real time scale simulators of high inherent accuracy will be possible. At present, exclusive of consideration of the time taken for setting up, programming, etc., certain problems which might be solved on a digital machine would take several hundred times longer in actual machine computation time than the same problem on the TRIDAC.[10]

High precision was thus thought less important than computing speed. The accuracy of results could after all be checked on a digital computer, and indeed from time to time check solutions were run on RAE's digital computer (an English Electric Co., DEUCE). However, because the digital check solution for a missile manoeuvring in three dimensions would take too long to compute, the problem had to be simplified for DEUCE. Yet even the simplified problem took 200 times as long to run on DEUCE as it took on TRIDAC.[11]

The TRIDAC facility enabled complete guided missile systems and target/missile interception scenarios to be simulated without the need to launch expensive prototypes. As we have already seen for the USA, work on the development of large electronic analogue computing facilities was not undertaken to further the state of the art in analogue computing *per se*. The primary motivation for their development was a combination of the technical demands and economic considerations arising from the design and testing of guided weapons. In 1957 Lt-Com F. R. J. Spearman, of the Ministry of Supply, summarised the combination of factors that was seen as justifying the construction of TRIDAC:

> To carry out experimental tests on a real system, as in the case of guided weapons, presents very formidable difficulties both from a practical and an economic aspect. ... The costs and uncertainties of full-scale trials are so high that the construction of elaborate and costly simulators is fully justified.[12]

It is difficult to estimate the extent of the influence of work on analogue computer development at, and for, the RAE on the design of commercial systems or on electronic analogue computation in general. One easily discernible consequence was the development of a small general-purpose electronic analogue computer, which Elliott developed at the behest of the MoS. The Elliott G-PAC (General-Purpose Analogue Computer) entered commercial production in 1955. Perhaps the most significant contribution of the work at the RAE was that the development and successes of GEPUS and TRIDAC helped to establish the utility of electronic analogue computers in aeronautical design. The work there also helped to establish computer simulation as a viable method for significantly reducing the cost of developing and testing new aircraft and guided weapons. In these respects the RAE and TRIDAC played a very similar role to the ONR's projects Cyclone and Typhoon in the USA.

THE EMERGENCE OF COMMERCIAL GENERAL-PURPOSE ELECTRONIC ANALOGUE COMPUTERS, 1945–1960

For Britain as well as the USA (see chapter 5) the firms in the electronics and aeronautics industries, and other organisations involved in the development and/or commercialisation of general-purpose electronic analogue computers, can be divided into three categories: the developer/users, the user/manufacturers and the non-user/manufacturers.

Examples of developer/users in Britain were the Sperry Gyroscope Company, A. V. Roe Ltd, Vickers-Armstrong Ltd, and the General Electric Company. These firms developed and built general-purpose electronic analogue computer systems for in-house use rather than as commercial products. Firms in the user/manufacturer category also developed analogue computers primarily for in-house use, but subsequently diversified and began to manufacture them on a commercial basis. In Britain by 1958 there were several user/manufacturer firms: Short Brothers and Harland Ltd, Saunders-Roe Ltd, English Electric Co and Fairey Aviation Ltd. The third category refers to those firms which developed and built analogue computers, not for their own use but primarily as a commercial product. For Britain, Elliott Brothers (London) Ltd and the electronics firm Solartron Ltd are the only two firms in this category.

In the following two sections I will first look at the firms in the aeronautics industry, and in the second section at the firms in the electronics industry.

Electronic Analogue computers and the British aeronautics industry

Short Brothers and Harland of Belfast, a leading manufacturer of aircraft, was the first British firm to introduce a commercial general-purpose electronic analogue computer. In 1953, Shorts began to advertise the availability of their new DC electronic analogue computer in trade magazines and engineering journals. Shorts' computer was a repetitive-operation system. It was clearly influenced by repetitive operation electronic analogue computing components developed by the American firm George A. Philbrick Researches Inc. Shorts even used Philbrick's convention of spelling computer, "computor".[13] Shorts' involvement in the development of electronic analogue computers originated from the increasing demand for aids to computation from its aircraft design team. Commenting in the early 1950s on the lack of suitable computing facilities, one of Shorts' design engineers, E. Lloyd Thomas, noted that the situation had brought about a transfer of attention "from the problem of formulation to that of solution".[14] Consequently, Shorts became a developer/user of electronic analogue computers, but soon decided to commercialise their "computor" and in doing so made the transition to user/manufacturer.

Shorts' customer list for the mid-1950s indicates that demand for the new computer came from many different types of organisation, and they were sold not only in Britain but also in other European countries (see Table 6.1). Yet by 1956 Shorts had only sold an esti-

TABLE 6.1 LIST OF CUSTOMERS FOR THE SHORTS GENERAL-PURPOSE ELECTRONIC ANALOGUE COMPUTER, 1955[15]

Admiralty Gunnery Establishment, Portland, Dorset.	Metropolitan-Vickers Electrical Co. Ltd
Armament Research and Development Establishment, Fort Halstead	Queen Mary College, London
	Royal Aircraft Establishment, Farnborough
Brunswick Technical Institute	Ultra Electric Co. Ltd
De Havilland Engine Co. Ltd	Vickers-Armstrong (Submarines) Co. Ltd
English Electric Co. Ltd	
Folland Aircraft Co. Ltd	Fiat (Aviation Division), Turin
Handley Page Co. Ltd	Netherlands National Aeronautical Research
Institute	
King's College London	Stenhardt Ingeniorsfirma AB, Stockholm

mated 20 computer systems.[16] The chief explanatory factor that must be taken into account regarding the quite small number of machines sold, is that many of the potential customers had, or were developing, electronic analogue computers of their own. Also, at a cost of nearly £5,000 the computer was not inexpensive.

Within a year of Shorts' entrance into commercial analogue computer manufacture, Saunders-Roe Ltd became the next user/developer firm from the aircraft industry to diversify in this direction. By the mid-1950s Saunders-Roe, like Shorts, had done a great deal of R&D in commercial electronic analogue computers.[17] They differed from Shorts in that initially they did not manufacture a standardised computer system. Instead, each system was tailored to meet the characteristics of the computational problems faced by individual customers. This involved developing a series of computing units which could be combined to create small to very large general or special-purpose computer systems. Saunders-Roe's first major customer for analogue computing equipment was the Admiralty Gunnery Establishment in Portland, Dorset.[18] However, contrary to their initial philosophy, Saunders-Roe did decide to begin manufacturing a small standard electronic analogue computer system, largely for use in preliminary design investigations and educational applications. Introduced in 1955, the "Miniputer" was a desk-top system, and having only ten computing amplifiers, it was a smaller machine than either the Shorts computer or the G-PAC which Elliott launched the same year.[19]

Whereas Shorts and Saunders-Roe had originally developed their electronic analogue computer equipment for use in the design of aircraft, English Electric (Guided Weapons) and Fairey Aviation developed analogue computers specifically as aids to computation in the design of guided weapons. The involvement of these firms in guided weapons development began in the late 1940s after the MoS realised that insufficient progress was being made in guided weapons development and that the task was beyond the resources available in the government's own research establishments. Thus, in 1948 the MoS set about encouraging private industry to participate. The English Electric Co. Ltd and the Fairey Aircraft Co. were among the first firms to be enrolled.[20]

The Guided Weapons Group at English Electric was established in 1949 but for security reasons was known as the Navigational Projects Division until 1954. Luton was chosen as the site for the new Division, where missile development activity was concealed under a programme known as the Navigational Control Project. The newly appointed Chief of the Guided Weapons Division was L. H. Bedford. During the war Bedford had worked at the Telecommunications Research Establishment on the development of analogue devices for anti-aircraft gun control. Consequently, much of the early work on analogue computing devices at English Electric (GWs) was based on Bedford's wartime experience.[21] At Luton, Bedford's design team began by developing a number of small special-purpose analogue computing devices. In the early 1950s R. W. Williams was a leading member of Bedford's design team at Luton. In 1968 Williams, then Professor of Electrical Engineering at Bath University, noted in a retrospective:

> The design and construction of [guided weapon] simulators was, in those days, necessarily a "do-it-yourself" activity since no standard [analogue] computing units were available on the British market. [However] with the passage of time, the limitations of the special-purpose approach became more and more apparent.[22]

The growing awareness of the inherent inflexibility of special-purpose analogue devices led to the design and construction of

English Electric's first general-purpose DC electronic analogue computer between 1953 and 1954. This was the predecessor of English Electric's commercial computer known as LACE (Luton Analogue Computing Engine). By 1956, all of the older special-purpose analogue computing equipment for missile modelling and simulation had been replaced by LACE computer systems.[23] The first public demonstration of the LACE (Mk II) computing system was not until 1958. The occasion was the first British Computer Exhibition, at Olympia in London, from 24 November to 4 December.[24] Also on display was a general-purpose electronic analogue computer, known as FACE (Fairey Analogue Computing Engine), manufactured by the Fairey Aviation Company. In the mid-1950s Fairey developed a range of general-purpose analogue computing equipment, primarily for guided missile development, at its factory in Heston, Middlesex.[25]

Though English Electric (GWs) and Fairey were among the earliest firms to undertake work on the development of general-purpose electronic analogue computers, they were also among the last of the developer/user firms from the aircraft and guided weapons industry to diversify into manufacturing analogue computers as a commercial product. Indeed, in 1958 the commercialisation of the English Electric LACE had only recently been completed. Prior to it entering commercial production, responsibility for the manufacture of LACE was transferred from the Guided Weapons Division at Luton to English Electric's Control and Electronics Department at Kidsgrove. There, it was redesigned and repackaged for commercial manufacture and renamed the LACE Mk II.

By 1958, four firms from the aeronautics industry had diversified into the commercial manufacture of general-purpose electronic analogue computing systems. In doing so, all four firms—Shorts Saunders-Roe, English Electric (GWs) and Fairey Aviation—had made the transition from developer/user to user/manufacturer. A number of firms in the aeronautics industry developed and used electronic analogue computing systems, but did not manufacture them on a commercial basis. For example, in the mid-1950s Vickers-Armstrong (Aircraft) Ltd built a general-purpose electronic analogue computer at their factory in Weybridge, Surrey. In the late 1950s staff at De Havilland Propellers Ltd's Simulation Laboratory designed and built one of Britain's largest centrally programmed electronic analogue computing facilities.

Electronic analogue computers and the British electronics industry

In Britain between 1945 and 1960 many firms in the electrical and electronics industries also developed special and general-purpose electronic analogue computers. As in the aeronautics industry, there were developer/user and user/manufacturer firms.

One of the first developer/user firms in Britain was the Sperry Gyroscope Co. Ltd. In the late 1940s Sperry built a general-purpose electronic analogue computer for in-house use. Sperry's "Electronic Simulator D.A. Mk1" was exhibited for the first time in 1949, at the Physical Society's Thirty-third Annual Exhibition of Scientific Instruments and Apparatus in London. Sperry was a subsidiary of the American firm of the same name, which specialised in navigation and ordnance fire-control equipment. The Mk1 was undoubtedly influenced by the rapid growth of interest in the use of electronics in analogue computer design in the USA. Indeed, in the USA Sperry's parent company was also working on the development of a similar computer. Though several Mk1s were built by Sperry (UK) for in-house use, the computer was never commercialised, nor was the American system.[26]

Another example of a developer/user firm is the General Electric Company Ltd (GEC). In the mid-1950s GEC developed "Stardac", a general-purpose electronic analogue computer solely for in-house use. It was designed and built by staff from GEC's Applied Electronics Laboratories in Stanmore, Middlesex, to solve problems associated with the development of guidance and control systems for missiles. Stardac was a large system, consisting of 200 computing amplifiers, and was the first British system to be operated from a central control desk.[27] The EMI Electronics Company, on the other hand, is an example of a firm from the electronics industry that developed, used and subsequently manufactured general-purpose electronic analogue computers commercially.

The first British non-user/manufacturer of general-purpose electronic analogue computers was Elliott Brothers (London) Ltd.[28] Elliott's involvement in electronic analogue computer development began, as we have seen, in the early 1950s with the design and construction of the TRIDAC computer system for RAE, Farnborough. Neither TRIDAC, nor the firm's first small commercial analogue computer—the Elliott G-PAC (General-Purpose Analogue Computer)— were developed for in-house use. Instead, both were projects undertaken at the behest of, and with funding from, the Ministry of Supply. G-PAC was designed under a Ministry of Supply development contract to provide a commercial general-purpose machine that would be inexpensive and easy to operate.[29] Elliott began manufacturing the G-PAC in 1955. Its selling price was £3,800, which was £1,200 less than the Shorts electronic analogue computer.

Whether or not Elliott would have developed a small commercial analogue computer without funding from the Ministry of Supply is a moot point. However, Elliott's involvement in electronic analogue computer development continued, and by 1958 Elliott was manufacturing two general-purpose systems; the G-PAC, which contained 20 computing amplifiers, and the Miniac, a desk-top computer with 10 computing amplifiers. Elliott also developed a special-purpose computer known as FIMAC (Fighter Interception Manoeuvres Analogue Computer) at the firm's own Aviation Division.[30]

Solartron began manufacturing general-purpose electronic analogue computers in 1957. By the early 1960s it was one of Britain's leading producers of electronic analogue computers, and it was the first firm of British origin to develop a commercially successful range of hybrid computers. Founded in 1947, Solartron's first product to include a small electronic analogue computer was a radar simulator introduced in 1955. Following the success of the simulator, Solartron began marketing a range of individual DC operational amplifiers and power supplies that could be used to build electronic analogue computing systems. In 1957 Solartron launched the "Analogue Tutor". Popular with colleges and universities, the small (5-amplifier) low-accuracy machine was used primarily for teaching purposes.

In 1958 Solartron developed a more versatile high-precision computer, the 10-amplifier SC 10 or "Minispace". By June 1961 forty Minispace computers had been sold, each costing approximately £1,600. Also in 1958, Solartron produced their first two SC 30s, both of which were bought in advance by the British Thomas Houston company. The SC 30, a 30-amplifier machine costing approximately £10,000, established Solartron as one of the leading firms among the companies that were manufacturers of high-quality, high-precision analogue computers for professional use. By mid-1961 more than a dozen SC 30s had been sold, about half going to foreign customers. The SC 30 saw a marked departure from the traditional upright electronic component racks used by most other British manufacturers. Solartron chose to design the computer in the form of a desk type console similar to contemporary digital computer control stations. In their advertisements they encouraged an office rather than an engineering laboratory setting for their machine.[31] By the early 1960s Solartron was manufacturing a wide variety of electronic analogue computers for educational and industrial use.[32] Moreover, Solartron was the only firm of British origin to continue to develop and manufacture its own range of commercial electronic analogue and hybrid computer equipment after 1965.

Several other firms from the electrical and electronics industry began manufacturing special and general-purpose electronic analogue computers in Britain in the mid-to-late-1950s.[33] Of these, the most notable were: EMI Ltd, Metropolitan-Vickers Ltd and Redifon Ltd. EMI's first commercial electronic analogue computer went on display at Britain's first computer exhibition at Olympia in 1958. Known as EMIAC II (EMI Analogue Computer), it contained 18 computing amplifiers, and like most general-purpose electronic analogue computers it was modular, so that several units could be coupled together for larger problems. In the late 1950s and early 1960s EMI had a measure of success, selling EMIACs to engineering departments in colleges and universities. But the machine was less popular among industrial users. In the 1960s EMI did not pursue the development of hybrid computers and by 1967 had stopped analogue computer manufacture altogether.[34]

In 1958, Metropolitan-Vickers was also manufacturing a general-purpose electronic analogue computer, the MV 952. Metropolitan-Vickers was a subsidiary of the Associated Electrical Industries Ltd (AEI) group of companies, and in 1960 the MV 952 was superseded by the AEI 955 analogue computer.[35] However, AEI's analogue computer products were not well known or seen as competitive with other commercial systems in the early 1960s. By 1967 AEI had also left the analogue computer business.[36]

In 1960 Redifon Ltd, a member of the Rediffusion Group and a leading manufacturer of flight simulators and air trainers, also began manufacturing general-purpose electronic analogue computer systems. The RADIC (Redifon Digital Analogue Computing System) was the first commercial "hybrid" computer developed by a British firm. The RADIC combined a 48-amplifier electronic analogue computer and a small parallel digital computing system for control, data input/output, conversion and storage.[37] Though it had digital control and storage, the analogue section of the RADIC did not compare favourably with the latest systems from American firms such as Electronics Associates Inc.[38] Redifon had few customers for the RADIC analogue/digital system. By 1963 Redifon's Computer Division was instead manufacturing two rather more conventional general-purpose electronic analogue computer systems; the Redifon 10/20 Computer, a 48-amplifier machine, and the Redifon 2/64 Computer, a 128-amplifier machine, both 0.1% accuracy systems. These computers were more popular with colleges and other educational establishments than with industrial users.[39] Competition from American firms led Redifon, in 1965, to discontinue its own research and development in electronic analogue computers. However, the firm did not leave the industry entirely. Redifon entered into an agreement with an American firm, Comcor Inc., a subsidiary of Astrodata Inc., to assemble and sell systems designed in the USA by Comcor.

Electronic analogue computers and the electricity-generating and nuclear industries

In addition to their widespread use and development in the aeronautics and electronics industries, electronic analogue computers were also developed and widely used by firms in the nuclear power and electricity-generating industries. Both general and special-purpose systems were developed. The special-purpose systems tended to be simulators, designed to train personnel in safety procedures and to test the efficacy of reactor control systems. General-purpose electronic analogue computers were used in the design and development of nuclear reactors.[40]

In Britain, Vickers-Armstrong (South Marston) Ltd, General Precision Ltd and Miles Highvolt Ltd were among the numerous companies that built reactor simulators for the nuclear industry and other research establishments. Also, several of the firms manufacturing general-purpose analogue computers either modified existing equipment or designed one-off systems for this purpose. For example, soon after the launch of Shorts' new SIMLAC computers in 1959 Shorts modified and sold one to the Operations School at the Calder Hall nuclear power station. Elliott and English Electric also manufactured general-purpose electronic analogue computers and special systems for nuclear reactor studies.

Elliott became involved in the construction of electronic analogue computing equipment for the nuclear industry in the mid-1950s, when it undertook to build a special-purpose electronic analogue computer for GEC's Simon-Carver Laboratories in Erith. Installed in 1956, it was capable of simulating the kinetic behaviour of an entire nuclear power station. The simulator was used in problems associated with gas-cooled graphite-moderated reactor stations and was reportedly the most comprehensive power station simulator facility in Britain.[41]

By the late 1950s English Electric had grown into a highly diversified company with interests in aerospace, nuclear and electrical

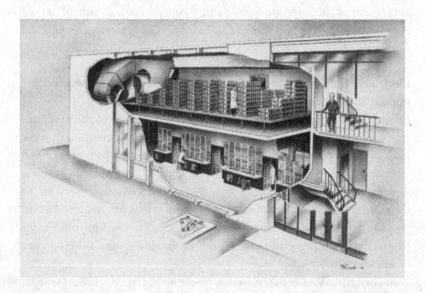

FIGURE 6.1 ENGLISH ELECTRIC SATURN

THE MARS GENERAL PURPOSE ANALOGUE COMPUTER.

FIGURE 6.2 ENGLISH ELECTRIC MARS

engineering and digital computing. Though the Guided Weapons Division of English Electric undoubtedly led the way in the development of general-purpose analogue computing equipment, by 1958 the company's Atomic Power Division had also developed two commercial electronic analogue computers for nuclear reactor studies (see Table 6.2). Moreover, in the late 1950s staff from the Atomic Power Division, at Whetstone near Leicester, developed and built two large electronic analogue computers for in-house use.[42] This division of English Electric was involved in the design of large civil nuclear power stations. In the design of a nuclear reactor, information was needed in regions of operation where it would be too dangerous to operate the real system. Therefore, in 1957 English Electric took the decision to develop a large general-purpose electronic analogue computer. This would have to be able to deal adequately with the largest possible computation and simulation problems envisaged at that time and still allow for future developments.

The two computers built at Whetstone for this purpose were called Saturn and Mars. Saturn was the larger of the two. It took three years to build, occupied two floors of a purpose-built laboratory and contained 1,512 computing amplifiers. In fact, it was the largest

TABLE 6.2 British Electronic Analogue Computers, Manufacturers and Models, 1953 to 1960[43]

Manufacturer	Date Introduced	Computer Models	Description	Price
Elliott Brothers (London) Ltd	1955	G-PAC	Expandable General-Purpose Electronic analogue computer (GPEAC) 20 op-amps.	£3,800
	1958	FIMAC	Flight simulator, 80 op-amps.	£—
	1958	MINIAC	Desk-top GPEAC, 10 op-amps.	£900
EMI Electronics	1958	EMIAC II	Expandable GPEAC, 18 op-amps.	£10,000
English Electric (GWs)	1958	LACE Mk II	Expandable GPEAC, 12 op-amps.	£7,000
English Electric (APD)	1958	KAE I EERS	Special-purpose Nuclear Reactor Simulators, 16 and 6 op-amps.	£—
Fairey Aviation Co.	1958	FACE	Expandable GPEAC, up to 96 op-amps.	£—
Metropolitan Vickers Co	1958	MV 952	Expandable GPEAC	£—
Redifon Ltd	1960	RADIC	GPEAC with digital controller, 48 op-amps.	£—
Saunders-Roe Ltd	1954	SARO EAC units	Range of GPEAC units	£—
	1955	Minipunter	Desk-top GPEAC, 10 op-amps.	£1000
Short Brothers and Harland Ltd	1953	Shorts analogue computer	Repetitive operation, GPEAC, 18 op-amps.	£5,000
Solartron Ltd	1959	SIMLAC	Large GPEAC 112 op-amps.	£50,000
	1957	Analogue Tutor	Dest-top GPEAC, 10 op-amps.	£900
	1958	Minispace (SC10)	Desk-top GPEAC, 10 op-amps.	£1,600
	1958	SC 30	GPEAC 30 op-amps.	£10,000

FIGURE 6.3   SHORT BROTHERS AND HARLAND GENERAL-PURPOSE ANALOGUE COMPUTER

general-purpose electronic analogue computer installation in Britain in the early 1960s. Saturn could either be operated as a single machine or it could be divided into six 252 amplifier sections, each with its own control panel. It is worth noting that among the general-purpose systems available on the market in 1960, a 200-amplifier computer was considered a large machine. Yet Saturn was by no means under-utilised: by 1961, 80% of the machine was in full-time use. By 1962 the largest single problem to have been set up on the computer required nearly 400 amplifiers and involved inter-connecting two of Saturn's computing units. In general, however, problems required no more than 250 amplifiers and could be solved using one section of the Saturn computer. Mars was the second large

FIGURE 6.4 SAUNDERS-ROE GENERAL-PURPOSE ANALOGUE COMPUTERS

FIGURE 6.5 FAIREY AVIATION FACE

FIGURE 6.6 ENGLISH ELECTRIC (GWs) LACE

analogue computer built at Whetstone for nuclear reactor studies. Much smaller than Jupiter, Mars had 210 amplifiers.[44] Taking into account the combined R&D effort undertaken at the separate divisions of the firm the 1950s, English Electric was without doubt Britain's principal developer/user of electronic analogue computing equipment.

FIGURE 6.7 ELLIOTT BROTHERS G-PAC

FIGURE 6.8   SOLARTRON SC 30

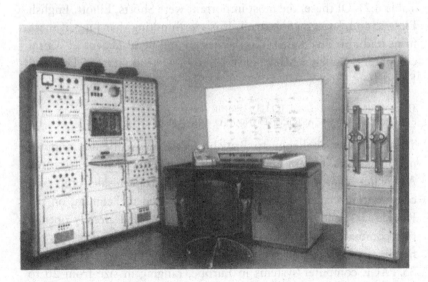

FIGURE 6.9   REDIFON RADIC

FIGURE 6.10   EAI PACE 231-R BEING INSTALLED AT THE UKAEA, WINFRITH HEATH, 1959

By the late 1960s nine major firms from the increasingly inter-twined British aeronautics and electronics industries were manufac-turing general-purpose electronic analogue computer systems (see Table 6.2). Of these, the most important were Shorts, Elliott, English Electric and, latterly, Solartron. It is worth noting that four of these firms had also diversified into the design and development of digital computers, namely EMI, English Electric, Metropolitan-Vickers and Elliott.[45]

THE AMERICAN CHALLENGE TO ELECTRONIC ANALOGUE COMPUTER FIRMS IN BRITAIN: ELECTRONIC ASSOCIATES INC./LTD, 1957–1962

The American challenge to British electronic analogue computer manufacturers began in earnest in July 1957, when Electronic Associates Incorporated established its first European sales office and computation centre. These offices were located in Belgium, on the Rue de la Science in Brussels. In the USA, EAI had already opened two commercial computing centres, but the Belgian centre was the first of its kind in Europe.[46] In the following two years EAI sold 35 PACE computer systems in Europe, ranging in size from 20 to 200 computing amplifiers.

EAI was the first American firm to make significant inroads into the British market for electronic analogue computers. However, initially it had great difficulties in doing so. This was largely because most of the potential customers in the aeronautics industry had already developed their own analogue computer, or had recently bought equipment from British firms. EAI's first customers were from the nuclear industry and government-funded research establishments working on nuclear research. In Britain, work on the development of electronic analogue computers for nuclear reactor studies had begun in the mid-1950s at the Atomic Energy Research Establishment of the United Kingdom Atomic Energy Authority (UKAEA) at Harwell. A machine containing 40 amplifiers was developed there for in-house use. However, computational problems soon outgrew the capacity of this system. In 1959 the analogue computer section, and much of the experimental work on nuclear reactors, was transferred from Harwell to the Atomic Energy Establishment (AEE) at Winfrith. To equip the new computer section at Winfrith, the UKAEA bought two EAI PACE 231-R (80-amplifier) systems at a total cost of more than £100,000. The Winfrith computers were imported from the USA and were the first EAI systems to be installed in Britain. In 1957 EAI also sold two PACE 231-R computers to the UKAEA for its research establishment at Risley, where similar studies to those at AEE Winfrith were in progress.

Also in 1957, PACE computing systems were purchased by the Associated Electrical Industries (AEI), by the Central Electricity Generating Board (two PACE 231-Rs) for their use at their headquarters in London, and by the Ministry of Supply for the Royal Armament Research and Development Establishment (RARDE) at Fort Halstead (one PACE 231-R).[47] In 1960 EAI established a subsidiary in Britain known as Electronics Associates Limited (EAL), located at Burgess Hill in Sussex. This was EAI's first manufacturing facility outside the USA and it provided the company with better access to the British and European markets. The first EAI computer (a PACE 231-R) assembled at Burgess Hill was purchased by the UKAEA in 1960 for its Atomic Weapons Research Establishment (AWRE) at Aldermaston. By the mid-1960s a second PACE 231-R had been installed at AWRE, Aldermaston, and the UKAEA had also purchased a PACE 221-R for the AWRE at Foulness. EAL also provided EAI with a British base from where staff could be deployed to meet the maintenance and support contracts which were a lucrative and important part of the analogue computer business. As part of their customer support activities EAL established a PACE User Group (PUG) in the summer of 1961, and opened a commercial computing centre at Burgess Hill.[48]

The bulk of the work at Burgess Hill consisted of the assembly and "anglicization" of PACE systems imported from the USA. In the mid-1960s, however, the engineers at Burgess Hill surprised the parent company by designing and developing a table-top medium-capacity system known as the EAL 380, which sold well in Britain. In the early 1960s the demand for EAL computers grew rapidly, and by the beginning of 1962 EAL had sold eighteen PACE 231-R computer systems, averaging 80 amplifiers each, and seven smaller PACE computer systems (see Table 6.3).

In the late 1950s and early 1960s British electronic analogue computers were markedly less advanced in terms of their technical specifications, user interfaces, programming aids and reliability than those manufactured by EAI/L. The technical superiority of EAI/L's computers was the basis for their success in the British market. EAL was particularly successful in the high-precision end of the British market for electronic analogue computers. By 1961 EAL was justifiably claiming to be responsible for "80% of all the electronic analogue computing systems in Britain having an accuracy of better than 0.1%".[49]

In the mid-1960s several other American firms established footholds in the British market, and they too dealt primarily in high-precision electronic analogue and hybrid computer systems. Of the British firms that began manufacturing analogue computers in the 1950s, Solartron, and to a much lesser extent Shorts and Redifon, were the only ones to present a serious commercial challenge to the American firms during the 1960s.

ELECTRONIC ANALOGUE AND HYBRID COMPUTING IN BRITAIN IN THE 1960S: THE AMERICAN ASCENDENCY

In the years between 1958 and 1963 the number of firms manufacturing general-purpose electronic analogue computers in Britain was at its greatest. The ten principal firms were AEI, EAL, Elliott, EMI, English Electric, Fairey, Redifon, Saunders-Roe, Shorts and Solartron. Of these, Solartron was the only firm of British origin manufacturing

TABLE 6.3 ELECTRONIC ASSOCIATES INC/LTD PACE COMPUTERS INSTALLATIONS IN BRITAIN IN 1962[50]

User	Computer Model	Quantity	No. of Computing Amplifiers
UKAEA, AEE, Winfrith	231-R	2	164
UKAEA, Risley	231-R	2	160
UKAEA, Atomic Weapons Research Establishment, Aldermaston	231-R	1	60
The Nuclear Power Group, Knutsford	231-R	2	200
Royal Armaments Research and Development Establishment, Fort Halstead	231-R	1	90
Admiralty Underwater Weapons Establishment, Portland	231-R	1	80
Bristol Aircraft Ltd, Filton	231-R	1	70
Folland Aircraft Ltd, Hamble	221-R	1	38
Vickers Armstrong (Aircraft) Ltd, Weybridge	231-R	1	100
Westland Aircraft Ltd, Yeovil	231-R	2	160
Central Electricity Generating Board, London.	231-R	2	205
General Electric Co., Erith	231-R	1	80
United Steel Co Ltd., Rotherham	221-R	1	32
Motor Industries Research Association, Nuneaton	231-R	1	48
Vauxhall Motors Ltd, Luton		1	—
Cambridge University	221-R	1	100
	231-R	1	—
Loughborough College of Advanced Technology	131-R	1	60
University College, Swansea	31-R	1	48
University of Leeds		1	—

electronic analogue computers in 1970. In addition to the competition from American analogue computer firms—and both British and American digital computer firms and their products—two factors were influential in the fairly rapid decline of the indigenous British analogue computer manufacturers. Firstly, the number of customers started to decline, largely because of important structural changes in the organisation of Britain's aeronautics industry. These changes began in the late 1950s at the behest of the British government. This restructuring led to the merger of the English Electric Aircraft Co., Ltd, Vickers-Armstrong (Aircraft) Ltd and the Bristol Aeroplane Co. to form the British Aircraft Corporation (BAC). In addition to these changes, a number of projects were cancelled. In 1962, for example, the British government cancelled the PT428 and Blue Water missile projects and as a result English Electric closed its Guided Weapons Division at Luton.[51] The following year, what remained of English Electric (Guided Weapons) merged with a similar division at the Bristol Aeroplane Co. to form a wholly owned subsidiary of BAC known as BAC (Guided Weapons) Ltd. Indeed, the goal, and overall effect, of the restructuring was to concentrate the aerospace industry so that there were fewer firms working on fewer projects. Firms which had in the 1950s diversified into the manufacture of electronic analogue computers were among those under pressure to rationalise, and therefore to consider their future involvement in computer development.

Secondly, financial and technical problems existed, largely associated with the development of higher-precision and higher-bandwidth computing components. Analogue computer manufacturers were aware that research and development costs rose sharply for systems with component accuracies above 0.01%. Yet in terms of funding for R&D, British firms were at a disadvantage compared to American firms, because of their lack of success in attempts to break into the larger and more lucrative North American market for electronic analogue computing equipment. By the mid-1960s the combination of competition from American firms, rationalisation in the aerospace industry and high development costs led AEI, Elliott, EMI, English Electric, Fairey and Saunders-Roe to discontinue the commercial manufacture of electronic analogue computers.

Nevertheless, throughout the 1960s there was still demand for new electronic analogue and hybrid computer equipment in Britain. In the mid-1960s former user/manufacturers and developer/users in the aerospace and electrical industries, as well as developer/users in research and educational establishments, began replacing older electronic analogue computer equipment. In contrast to the 1950s, these firms and organisations purchased computer systems instead of developing new ones internally.[52] Those firms which continued to manufacture electronic analogue computers, most notably Solartron, Shorts and EAL, benefited from the combination of a smaller number of manufacturers and the replacement demand from former user/manufacturers and developer/users.[53]

Shorts was one of the firms which made a serious attempt to compete for a share of the market for large high-precision systems. Because of rationalisation in the aircraft industry and the fall in demand for aircraft in the early 1960s, Shorts were keen to expand the analogue computer part of their business.[54] In the 1960s Shorts continued to develop their SIMLAC computer systems which they had launched in 1959. This was Shorts' high-precision machine and was aimed at professional use in research and industry. The SIMLAC cost approximately £50,000, contained 112 operational amplifiers, incorporated a patch board programming system and operated to a accuracy of 0.01%.[55] This system placed Shorts in direct competition with EAL in the high-precision, medium-to-large system end of the analogue computer market.

Nevertheless, in 1963 EAI claimed that its British subsidiary (EAL) was the sole manufacturer of high-precision analogue computer equipment in Britain. To support this claim, EAL introduced a classification scheme with Class I representing a computer with a minimum accuracy of 0.01%. EAL's claim that it was the only British manufacturer producing computers in this class did not go uncontested. In reply, H. G. Conway of Short Bros and Harland wrote:

> This is of course all wrong. ... we can take credit for a Class I machine (SIMLAC) which may well give the splendid American machine a run for its money. At present we count Babcock and Wilson, ICI, UKAEA and the MoA as customers with machines delivered.[56]

D. S. Terrett, EAL's Chief Engineer, responded for the company in a letter published in the journal *Engineering*. Terrett placed SIMLAC in Class II, which EAI had defined as the 0.1% accuracy category. He did so on the basis that to qualify for inclusion in Class I, all computing components had to operate within a 0.001% accuracy, and not only the linear computing amplifiers. On SIMLAC, the amplifiers met this specification, but the other components did not.[57] Indeed, Shorts implicitly acknowledged the superiority of the American equipment when in the mid-1960s it started buying electronic analogue computer systems from EAL. By 1967 Shorts had ceased all commercial production of electronic analogue computer equipment.[58]

Thus by the late 1960s there were just four major firms competing in the British market for high-performance electronic analogue and hybrid computers: EAL and Solartron, and two newcomers, Applied Dynamics Inc. and Redifon-Astrodata Ltd.

The Redifon-Astrodata partnership emerged in 1965 after the Computer Division of Redifon Ltd decided to stop designing and developing its own range of electronic analogue computers. It entered into an agreement with Astrodata of Anaheim, California, to manufacture and market the range of analogue/hybrid computer systems developed by Comcor Inc., a subsidiary of Astrodata. The new firm was called Redifon-Astrodata Ltd (RAL).[59] Among RAL's first customers were Rolls-Royce, the University of North Wales and RAE, Farnborough. The RAE bought two large Ci 5000 hybrid computer systems to replace the massive TRIDAC computer which had been in operation since 1954. However, RAL's computers did not meet the operational requirements of the RAE's guided weapons division. As a result, in 1967 the RAE sold both Ci 5000 computers to ICI, and instead bought two AD4 analogue computers manufactured by the American firm Applied Dynamics Inc.[60]

Applied Dynamics Inc. was the third American firm to establish itself in Britain. Yet unlike EAI/L and Redifon-Astrodata, Applied Dynamics chose not to open a manufacturing or assembly facility in Britain. Instead, Applied Dynamics chose a Dutch firm, Van Rietschoten and Houwens (R&H), as their European agents. R&H established a sales and support office in Croydon in the summer of 1966, and by 1968 had sold large electronic analogue computer systems to the Cranfield College of Aeronautics, Manchester University and RAE, Farnborough.[61]

Solartron launched its first comprehensive range of high-performance electronic analogue computers, the Solartron 247 Series,

TABLE 6.4 USERS OF SOLARTRON ELECTRONIC ANALOGUE/HYBRID COMPUTERS IN THE 1960s[62]

User	Analogue/Hybrid Computer	Digital Computer
University of Bath	Solartron 256/	DEC PDP 8
University of Swansea	Solartron 256/	DEC PDP 8
Salford Technical College	Solartron HS7/	DEC PDP 8
AEI, Leicester	Solartron HS7-6	
Atomic Energy Research Establishment, Harwell	Solartron HS7-6, × 2	
BAC, Preston	Solartron HS7-6, × 2	
ICI Billingham/Stockton	Solartron 247 (Largest of Solartron's analogue computer systems in the UK 1960–1965 system cost approx. £90,000)	
Lanchester College of Technology	Solartron 247	
UMIST, Manchester	Solartron 247	
English Electric, Nelson Labs, Luton.	Solartron 247	
National Physical Laboratory, Teddington	Solartron 247	
AEI, Trafford Park, Manchester	Space 30 × 2	

FIGURE 6.11   SOLARTRON H7 SERIES 1

in 1962. Between 1962 and 1964, 18 Solartron 247s were sold, and orders for 2 more had been received.[63] The Solartron 247 was followed in the mid-1960s by a range of hybrid computer systems known as the Solartron Hybrid 7 Series (H7S).[64] The computer series was fully transistorised and ranged in size and price from the H7S-1, a 24-amplifier system which cost less than £5,000, to the H7S-7, a 144-amplifier system which cost approximately £50,000.[65] Solartron's hybrid computers kept them in competition with the leading American firms. With these new systems, Solartron was able to maintain its position in the British market. In the late 1960s the firm's analogue and hybrid computer division was doing approximately

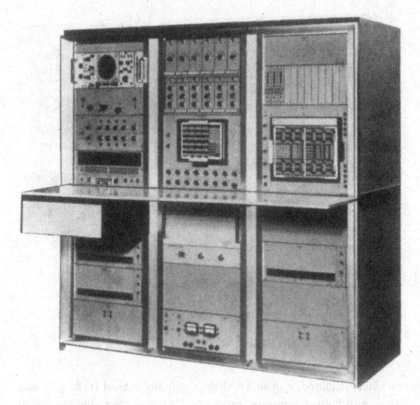

FIGURE 6.12 SOLARTRON H7 SERIES 3

£500,000 worth of business annually and, though some way behind, remained second only to EAL.[66] From 1965 to 1970 Solartron was the only firm offering a range of electronic analogue and hybrid computers designed and built entirely in Britain.[67] In Table 6.4 a list of some of Solartron's customers in the 1960s is given.

Yet, though Solartron was relatively successful in Britain, the firm was unable to establish a niche for itself in the analogue computer market in the USA. Moreover, by 1968 EAL alone accounted for approximately three quarters of the British market (estimated to be worth between £1m and £2m annually) for electronic analogue and hybrid computing equipment.[68] Despite attempts by Solartron to modernise its H7 Series, its product range was considered technically inferior to the latest high-precision and highly integrated hybrid computer systems available from US firms. Demand for its computer

FIGURE 6.13   *SOLARTRON H7 SERIES 7*

products declined, and in 1970 the company ceased trading in analogue and hybrid computer products. EAL was then able to consolidate its lead in Britain. In 1971 EAL paid Solartron approximately £90,000 for the firm's outstanding support and maintenance contracts and inventory of computer components. The dominance of EAI/L and Applied Dynamics in Britain also proved too great for Redifon-Astrodata, and in 1972 the firm stopped supplying electronic analogue and hybrid computer systems.[69]

HYBRID COMPUTING IN BRITAIN: THE ROLE OF USERS AND ECONOMIC ANALYSIS

In Britain during the 1960s there was a gradual shift away from all-electronic analogue computing towards hybrid computing. This trend was evident both in industry and in educational institutions. In the latter, hybrid computers served two main functions: the first was to provide the various engineering and science departments with a computing resource for their own research activities. The second was the provision of a computing facility that could be used to teach undergraduate students various aspects of control theory, mathematical modelling and simulation, and the solution of differential equations in many engineering applications. In practice, the availability of commercial systems meant that they were regarded more as a means to an end than as a subject for further research.

One of the chief "ends" to which hybrid computers were applied in British academic research was the field of control engineering. In the mid-1960s three "centres of excellence" in control engineering were selected for special support by the Science Research Council. These were at Cambridge University, Imperial College, London, and the University of Manchester Institute of Science and Technology (UMIST). The first to install a large hybrid computer was the Control Group, headed by Professor John Coals, at Cambridge University. The computer was purchased with a Science Research Council grant for £273,250 awarded in July 1965. Elliott-Automation Ltd (formerly Elliott Brothers) won the contract to build the hybrid computer system. It was completed and installed in the Control Laboratory at Cambridge University towards the end of 1967. It comprised two EAI PACE 231-R Mk V electronic analogue computers and an Elliott 4130 digital computer. It was fully operational by the following April and was at that time the largest hybrid computer installation in Britain.[70]

The Cambridge hybrid computer was soon followed by systems at UMIST and Imperial College. The UMIST hybrid computer was installed at the Control Systems Centre in 1968 but was soon being expanded. In 1970 the UMIST system consisted of three RAL Ci 175 electronic analogue computers and a Digital Equipment Corporation PDP-10.[71] The hybrid computer facility at Imperial College consisted of an EAI 680 electronic analogue computer and an EAI 640 digital computer. In the early 1970s the EAI 640 was replaced by an EAI PACER 1000 digital computer.

In the late 1960s the transition from analogue to hybrid computing was also taking place in industrial research and development laboratories and government research establishments. One of the first experimental systems was installed at English Electric's Nelson Research Laboratories in the mid-1960s. It consisted of a Solartron 247 analogue computer and a Marconi Myriad digital computer.[72] At RAE, Farnborough, hybrid computing superseded analogue. By the early 1970s there were two hybrid computing facilities at the

RAE. Each consisted of two Applied Dynamic AD 4 analogue computers linked to an IBM 1130 digital computer. In the aerospace industry, Shorts of Belfast replaced electronic analogue computing equipment of their own design with a hybrid computer installation consisting of a RAL Ci 5000 analogue computer and a Digital Equipment Corporation PDP 15 digital computer.

The British Aircraft Corporation, with its many divisions, was the largest owner of hybrid computing systems in Britain. BAC's first hybrid computer was an EAI 8945. It was installed at BAC's Guided Weapons Division at Stevenage in March 1968. It cost over £500,000 and was the first large hybrid computing facility in the British aerospace industry. In the late 1960s and 1970s large hybrid computing facilities were also in daily use at the BAC's Guided Weapons Divisions at Bristol and at its Military Aircraft Division at Warton. At BAC's Commercial Aircraft Division at Filton, hybrid and analogue computers were used in the design and development of the supersonic passenger aircraft, Concorde.[73]

However, in industry the continuing use of hybrid rather than all-digital computers for computation and simulation was subject to scrutiny and review. Industrial users in particular had to justify the use of hybrid computing techniques not only in terms of their technical merits, but increasingly in economic terms based on the cost per simulation/solution. One of the leading proponents of the use of economic analysis to determine, and justify, which computing technology should be used for a given problem was Michael Brown, the Head of the Hybrid Computing Group at BAC's Guided Weapons Division at Stevenage. The question on which he based his analysis was "Which computing tool gives the lowest cost per result to the accuracy I require?" In the mid-1960s Brown performed a comparison of all-analogue, hybrid and all-digital methods. He concluded that because the time to change between problems was very long on large all-analogue computers, they were only viable where real-time operation was essential and they could be virtually committed to one problem. He argued that for problems requiring a large number of runs, such as statistical studies, a large hybrid computer was much more economical than the all-digital computer. Table 6.5 gives details of the comparisons he made and the times and costs involved.

TABLE 6.5 COMPARISON OF ANALOGUE, DIGITAL AND HYBRID COMPUTING[74]

Factors	Analogue Computer	Digital Computer	Hybrid Computer
Time per run	10 sec.	300 sec.	2 sec.
Time per run allowing for parameter changes	120 sec.	300 sec.	3 sec.
Useful hours per shift	4 hrs.	8 hrs.	7 hrs.
No. of runs per shift	120	96	5,000 +
No. of shifts per day	1	3	2
Approximate cost per hour	£ 30	£ 70	£ 30
Approximate cost per run	£ 2	£ 6	£ 0.50
Approximate programming cost	£ 600	£ 300	£ 750
Approximate debugging cost	£4,800	£ 180	£ 720
Approximate re-entry cost	£2,400	£ 0	£ 15

TABLE 6.6 HYBRID COMPUTER INSTALLATIONS IN BRITAIN 1965–1975[75]

User	Analogue Computer	Digital Computer
Firms and Research Establishments		
CEGB	EAI 8800 ×3	EAI 8400
BAC, Stevenage	EAI 8800 ×2	EAI640/PACER100
BAC, Bristol	EAI 8800	DEC PDP 11/45
BAC, Warton	EAI 680 ×2	DEC PDP 11 + 15
Marconi Space & Defence Sys.	EAI 8800	EAI 8400
Rolls Royce & Associates	EAI 781	Sigma 5
Rolls Royce, Derby	EAI PACER 600	EAI PACER 600
Rolls Royce, Bristol	EAI PACER 600	EAI PACER 600
National Gas Turbine Estab.	EAI PACER 500	EAI PACER 500
Marine Industries Centre, University of Newcastle	EAI PACER 600	EAI PACER 600
Fords	EAI 680	IBM 1130
YARD, Glasgow	EAI 680	DEC PDP 15
Medical Research Council	EAL HY 48	Honeywell 516
RAE, Farnborough	AD 4 ×4	IBM 1130 ×2
RAE, Bedford	AD 4	Sigma 5
ICI	AD 4 ×2	Sigma 3
Short Bros and Harland	RAL Ci 5000	DEC PDP 15
BAC, Filton	EAI 680	
British Rail, Derby	EAI 680	
UKAEA, Winfrith	EAI 680	
AEI, Leicester	Solartron HS7-6	
Universities and Technical Colleges		
University of Bangor	RAL Ci 5000	HP 2116
Bath University	Solartron 256	DEC PDP 8
Cambridge University	EAI 231-R Mk-V ×4	ICL 4130
City University	EAI 680	EAI 640/ICL 1905
Cranfield Inst. of Tech.	AD 256	Micro 16/ICL 1905
Hatfield Polytechnic	EAI 680	DEC PDP 8
Imperial College	EAI 680	PACER 100
Salford University	Solartron HS7	DEC PDP 8
University of Swansea	Solartron 256	DEC PDP 8
UMIST	RAL Ci 175 ×3	DEC PDP 10
Liverpool Polytechnic	EAI 380 EAI 580	DEC PDP 8
University of Sussex	EAI 680	

Though BAC owned the greatest number of hybrid computers in Britain, the single largest hybrid computer installation was owned and operated by the Central Electricity Generating Board (CEGB).

This system cost approximately £1 million and was installed at the CEGB's national control centre at Southwark in 1969. It consisted of three EAI 8812 analogue computers and one EAI 8400 digital computer. The CEGB's decision to buy a hybrid computer rather than an all-digital system to replace its ageing EAI 231-R analogue computers followed more than three years of evaluation and investigation. During this time two parallel studies of the performance of all-digital computers and hybrid systems were carried out.[76] These studies indicated that considerable savings in computer time and overall costs could be made using the hybrid rather than an all-digital computing system.[77] Table 6.6 gives a list of the large hybrid computer installations in Britain, 1965–1975.

CONCLUSION

Thus, in Britain between the mid-1950s and the mid-1970s, electronic analogue computers had become well-established tools in industry and in engineering departments in universities and technical colleges. Their use continued into the late 1970s, but by that time the British market for electronic analogue and hybrid computers had been taken over by American firms. During the 1960s the role of developer/users and user/manufacturers declined, and the supply of electronic analogue and hybrid computers was left to specialised firms.

Compared with Britain, firms in the USA took an early lead in the development and commercialisation of general-purpose electronic analogue computers. Thus, in the late 1950s, the emerging British manufacturers—mainly user/manufacturers—had to compete with specialised American firms that were already well established in the USA. Whereas these American firms had considerable success at home and abroad, the British firms were less successful in entering foreign markets. In addition to competition from American firms, the British manufacturers were also faced with structural changes in the aerospace industry in the early 1960s. This restructuring reduced the single most important customer/user base. It also directly affected those manufacturers who had diversified from aeronautics into the commercial production of analogue computers. These firms were forced to consolidate and to withdraw from the manufacture of electronic analogue computer equipment. Those British firms that remained could not continue to meet the rising costs associated with the development of new high-precision electronic analogue and hybrid computers. From 1970 the demand for electronic analogue and hybrid computers in Britain was met entirely by firms of American origin.

NOTES

1. S. R. Twigge, *The Early Development of Guided Weapons in the United Kingdom*, Ph.D. thesis, University of Manchester, Manchester, 1990, p. 226, published as S. R. Twigge, *The Early Development of Guided Weapons in the United Kingdom, 1940–1960*, Harwood Academic Publishers, London, 1992.

2. The postwar reorganization which placed the RAE at the centre of guided weapons R&D in the UK was by no means RAE's first encounter with such programmes. During the war, two guided weapon projects were undertaken at the RAE. Projects Ben and Longshot were air-to-air missiles developed in cooperation with the TRE. Yet there is no evidence of any activity in electronic analogue computer development until after the 1945 reorganization took place. J. E. P. Gunning, "The Rocket Propulsion Establishment, Westcott," *Journal of the Royal Aeronautical Society*, Vol. 70, Jan 1966, pp. 285–287.

3. In April 1946 the newly elected Labour government reorganized the service departments, and the Ministry of Aircraft Production (MAP) and the Ministry of Supply (MoS) were merged to form the 'new' MoS.

4. The role of the GPE was downgraded and was reconstituted as the Rocket Propulsion Department of the RAE.

5. C. A. A. Wass, *Introduction to Electronic Analogue Computers*, Pergamon, London, 1955, p. ix.

6. J. J. Gait and D. W. Allen, "An Electronic Analyzer for Linear Differential Equations", *RAE Report: Guided Weapons*, 1, December 1947; *Catalogue of the Thirty-Second Annual Exhibition of Scientific Instruments and Apparatus*, The Physical Society, London, 1948; C. A. A. Wass, *Introduction to Electronic Analogue Computers*, p. 205.

7. M. Squires, "Interim Note on the Development of General Purpose Electronic Simulator", *RAE Technical Note: Guided Weapons*, No. 27, July 1948. GEPUS consisted of 30 units for scaling and setting initial conditions, 16 integrators, 15 summing amplifiers, 15 multipliers and 12 curve followers (or function generators). It was a fairly large system, occupying 15 (2 × 0.5 metre) laboratory racks. It contained some 850 thermionic valves and had a failure rate of approximately one per week. This relatively high system reliability was achieved, in part, by running each valve for several dozen hours in a valve-ageing panel prior to placing it in the machine. See also, C. A. A. Wass, *Introduction to Electronic Analogue Computers*, pp. 209–213.

8. Under peak operating conditions it absorbed 650 KW of electrical energy. Anon., "Three-Dimensional Analogue Computer", *The Engineer*, 15 Oct 1954, pp. 532–533. The MoS also commissioned Elliott Bros to build a smaller version of TRIDAC known as AGWAC (Australian Guided Weapons Analogue Computer). AWGAC was designed jointly by staff from Elliott Bros the MoS and the Department of Supply, Australia, and was installed at the Long Range Weapons Establishment, Salisbury, South Australia in 1954. See, P. Morton, *Fire Across the Desert*, Australian Government Publishing Service Press, Canberra, 1989, pp. 200–201.

9. Anon., "Three-Dimensional Analogue Computer: Predicting High-speed Flight", *Engineering*, 5 Nov 1954, pp. 596–597; F. R. J. Spearman, J. J. Gait, A. V. Hemingway and R. W. Hynes, "TRIDAC, A large analogue computing machine", *Proc., IEE*, Vol. 103, Pt. B, 1956, p. 375.

10. Anon., "Three-Dimensional Analogue Computer", *The Engineer*, 15 Oct 1954, p. 532.

11. Moreover, on the digital computer the result appeared in numerical form on cards which had to be converted into a graphical form. On TRIDAC the data were plotted as the calculation proceeded and was thus available immediately. However, whereas it took approximately the same time to program DEUCE as it took to set up TRIDAC for this type of problem, DEUCE had the advantage that it could be used for other problems during program preparation. Smaller systems were also used at the RAE for a variety of tasks and were seen as essential aids to design. Analogue computers were used in the design process to model isolated parts of the missile system, reduce uncertainties and optimise performance characteristics. Electronic analogue computers were also used in RAE departments outside of guided weapons development, for example in determining critical flutter speeds and analysing vibration problems in aircraft design. TRIDAC was designed as a "research tool for the applied physicist, aerodynamicist and engineer" who often worked with a great deal of uncertainty as regards the accuracy of data and the validity of extant theories of aerodynamics. Thus, where data were not generally known to a precision of better than 1%, the 0.1 % accuracy of TRIDAC solutions was not considered a serious limitation, and was far outweighed by the benefits of high computing speeds. Anon., "Three-Dimensional Analogue Computer", *The Engineer*, p. 532; F. Smith, "The Electronic Simulator for the Solution of Flutter and Vibration Problems", *RAE Report: Structures*, No. 51, October 1949; J. J. Gait, "Analogue Computers and their Applications", *Process Control and Automation*, Nov 1958, pp. 485–480; J. J. Gait, "Tridac: A Large Analogue Computer for Flight Simulation", *Journee International de Calcul Analogique*, Sept 1955.

12. F. R. J. Spearman, (Lt-Com) "Analogue Computing Applied to Guided Weapons," *Discovery*, May 1957, pp. 192–197.

13. E. L. Thomas, "A New Analogue Computor," *Engineering*, Vol. 176, Oct 1953, pp. 477–479; Anon., "A General Purpose Electronic Analogue Computer", *The Engineer*, 25 Sept 1953, pp. 395–397.

14. E. L. Thomas, "A New Analogue Computor," p. 477; R. J. A. Paul, "The SHORT Electronic Analogue Computer", *The Overseas Engineer*, Vol. 29, 1956, pp. 205–208; R. J. A. Paul and E. L. Thomas, "The Design and Applications of a General-Purpose Analogue Computer", *Journal of the British IRE*, Vol. 17, Jan 1957, pp. 49–73.

15. *Shorts General Purpose Analogue Computor*, Trade Publication, Short Bros and Harland, Belfast, c. 1955.

16. A. J. Knight, "A Survey of Computing Facilities in the UK", (2nd ed.), *Directorate of Weapons Research Report No. 5/56*, Ministry of Supply, London, August 1956, p. 41.

17. Anon., "The Aircraft Industry Display: Computers" *Engineering*, Vol, 180, 30 Sept 1955, pp. 471–474, p. 472.

18. A. J. Knight, "A Survey of Computing Facilities in the UK", p. 41.

19. Anon., "Miniature Analogue Computer", *Engineering*, Vol. 179, May 1955, p. 659.

20. In 1949 the MoS placed a Ministry Study Contract with English Electric for the investigation of the design of a ground-to-air missile. In August 1950 the contract was formalized and the project was given the name Red Heathen. This missile was intended primarily for use by the British Army and was the predecessor of the Thunderbird tactical nuclear missile. S. Twigge, *The Early Development of Guided Weapons in the United Kingdom*, Ph.D. thesis, p. 381–391.

21. I. N. Cartmell and R. W. Williams, "Guided Weapons Simulators", *Journal of the Royal Aeronautical Society*, Vol. 72, April 1968, pp. 356–360.

22. Ibid. p. 357.

23. Ibid. p. 357; R. J. Gomperts and D. W. Righton, "LACE: The Luton Analogue Computing Engine, Pt 1", *Electronic Engineering*, Vol. 29, July 1957, pp. 306–312; J. C. Jones and D. Readshaw, "LACE: The Luton Analogue Computing Engine: Pt 2", *Electronic Engineering*, Vol. 29, July 1957, pp. 306–312. As the LACE computer facility at Luton

continued to grow, many improvements were made and in 1958 the LACE Mark II was introduced. By the mid-1960s the computing facility contained approximately 440 computing amplifiers and over a hundred multipliers. As before with LACE and LACE Mk II, the development of new equipment over the period 1958 to 1966 was performed in-house. Rationalization and mergers in the British aerospace industry in the early 1960s resulted in the transfer of guided weapons development from English Electric, Luton, to BAC, Stevenage, and the LACE computing facility followed. However, with the increasing complexity of missiles systems and electronic models it was becoming clear that the LACE computer facility required modernising. To significantly improve the LACE II facility at Stevenage would have required BAC to fund a comprehensive R&D programme to develop new hybrid systems to match the technical specifications achieved by the latest American computer products. Analogue computer development at English Electric and BAC had been a means to an end much more than a core activity. In-house analogue computing development at the BAC, Stevenage, was therefore discontinued. From 1966 onwards the LACE computers were gradually replaced by systems manufactured by EAI in the USA. I. N. Cartmell and R. W. Williams, "Guided Weapons Simulators"; Personal Interview: Robin Penforld, BAe, Stevenage, 22 March 1991; Personal Interview, Barry J. Thompson: BAe, Stevenage, 22 March 1991.

24.   Four firms from the electrical and electronics industry also exhibited commercial analogue computer systems at Olympia, namely, Elliott Brothers (London) Ltd, EMI Ltd, English Electric Co., and Solartron Electronic Group Ltd. Anon., "An Automation Progress Survey: Principal British Analogue Computers", *Automation Progress* Nov 1958, pp. 403–405; Anon., "Electronic Computer Exhibition", *Process Control and Automation*, Vol. 5, no. 11, Nov 1958, pp. 487–496.

25.   G. E. Thomas, "Analogue Computers as Design Tools', *Automation Progress*, Vol. 3, Nov 1958, pp. 415–429; *Fairey Multi-purpose Analogue Computers*, Trade Publication, Fairey Aviation, Heston, c. 1958. From 1950 to 1957 Fairey were involved in the development of the Fireflash (formerly Blue Sky) air-to-air guided missile which entered service for the RAF in 1957. See, S. Twigge, *The Early Development of Guided Weapons in the United Kingdom*, Ph. D. thesis, p. 83.

26.   The Mk1 was a repetitive operation DC analogue computer. On the Sperry machine the entire calculation could be repeated at a frequency of between 0.5 and 50 times a second. C. A. A. Wass, *Introduction to Electronic Analogue Computers*, p. 205.

27.   P. L. Hodges, "Stardac—A General Purpose Analogue Computer: Part I", *Report No. SL.196*, General Electric Company, Applied Electronics Laboratory, 1 Feb 1957; E. Brown and P. M. Walker, "Stardac—A General Purpose Analogue Computer: Part II", *Report No. SL.202*, General Electric Company, Applied Electronics Laboratory,1 April 1957.

28.   Elliott Brothers (London) Ltd was also one of the first British firms to develop commercial digital computers: the Elliott 400 series. Established in 1801 the firm originally manufactured scientific apparatus, but during the Second World War Elliott was also involved in the development of electro-mechanical gun directors for the British Navy. S. Lavington, *Early British Computers*, Manchester University Press, Manchester, 1980, p. 57–61.

29.   Anon., "The Aircraft Industry Display: Computers", p. 472.

30.   Anon., "British Computers Show their Paces", p. 409. One of Elliott's customers for the G-PAC was the City University. London, which until May 1966 was known as the Northampton Polytechnic. Their G-PAC was purchased in 1958 and remained in operation until 1967 when it was replaced by equipment manufactured by EAI/L. Anon., "Computers and Computing at the City University", *Typescript Report*, City University, London, Sept 1981.

31.   Anon., "Solartron SC 30 Analogue Computer," *Engineering*, Vol. 191, June 1961, pp. 874–876, p. 876.

32. Solartron was one of the first British firms to open a commercial computing centre. This was located at their Dorking Research and Development Laboratories. The company grew rapidly: in the first five years of trading, exports amounted to £15,000, between 1955 and 1960 this rose to £1,181,000. In 1960 the company had three factories located at Thames-Ditton, Dorking and Farnborough, employing 1,330 people. In 1960 Firth Cleveland took control of the Solartron Electronic Group. The takeover was necessary in order to finance R&D for new product development. Anon., "Firth Cleveland Back Solartron Group," *Engineering*, Vol. 15. Jan 1960. p. 89.

33. In Britain two small companies—Cape Electronics in the late 1950s and Lan Electronics in the early 1960s—designed and built small, low-cost, general-purpose machines for educational use. Computer Consultants, *The European Computer Users Handbook*, 5th Edition, Computer Consultants Ltd, Enfield, 1967, section 7.

34. In the late 1950s and early 1960s EMIAC II cost between £2,000 and £3,000 depending on the selection and specification of particular sub-units. One of EMI's customers was Oxford University which in 1962 bought an EMIAC II for use in a newly opened engineering laboratory. The computer was to be used to analyze the behaviour of a variety of self-adaptive control systems. Anon., "Oxbridge Computes with the Times," *Engineering*, 8 June, 1962, p. 761.

35. The AEI 955 was manufactured at AEI Electronic Apparatus Division, Trafford Park, Manchester. Work on the design and development of analogue and small hybrid computer systems was also undertaken at AEI's Electronic Apparatus Division, Leicester in connection with the Bloodhound I and II missiles. *AEI 955 Analogue Computer*, Trade Publication, Associated Electrical Industries, Trafford Park, Manchester, May 1960; E. L. Thomas, "Analogue Computation", *British Communications and Electronics*, May 1958, pp. 348-358.

36. Computer Consultants, *The European Computer Users Handbook*, p. 3.

37. *RADIC: Redifon Analogue Digital Computing System*, Trade Publication, Redifon Ltd, Crawley, c. 1960; Anon., "Applications of the RADIC System", *Typescript Report*, Redifon Ltd, Crawley, c. 1960, pp. 1-7.

38. RADIC operated to an overall accuracy of 0.1% compared with the American computers which operated to 0.01%.

39. In 1963 Redifon's customers included The London Polytechnic, Maths Dept., Wolverhampton and Staffordshire College of Technology, Maths Dept., Southampton College of Technology, Electrical Engineering Dept., the Birmingham College of Advanced Technology, Mechanical Engineering Dept. *Redifon, Computer Department News Letter*, Redifon Ltd, Crawley, No. 3, 1963, pp. 1-4.

40. Indeed, in the 1950s analogue computers played an increasing role in the training of nuclear engineers and reactor operators and in the study of reactor physics. Anon., "Reactor Simulators," *Engineering*, 23 June, 1961, p. 864; G. J. R. MacLusky, "An Analogue Computer for Nuclear Power Studies," *Proc, IEE*, Vol. 104, Pt. B, no. 17 Sept 1957, pp. 433-442.

41. Anon., "Atomic Review: Pattern of Research", *Engineering*, Vol. 182, Aug 1956, pp. 283-285.

42. T. O. Jeffries, C. P. Newport, H. A. Darker and R. A. Flint, "A Large Analogue Computing Installation", *Nuclear Power*, Dec 1961, pp. 80-84; T. O. Jeffries, C. P. Newport, H. A. Darker and R. A. Flint, "The Saturn Analogue Computer", *Nuclear Power*, Jan 1962, pp. 69-72.

43. Anon., "Principal British Analogue Computers", *Automation Progress*, Nov 1958, pp. 404-405; Anon., "Miniature Analogue Computer", *Engineering*, Vol. 179, May 1955, p. 659; Anon., "The Aircraft Industry Display: Computers" *Engineering*, Vol, 180, 30 Sept 1955, pp. 471-474, p. 472; E. L. Thomas, "A New Analogue Computor," *Engineering*, Vol. 176, Oct 1953, pp. 477-479; E. L. Thomas, "Analogue Computation", *British Communications and Electronics*, May 1958, pp. 348-358; *RADIC: Redifon Analogue Digital Computing System*; *Applications of the RADIC System*.

44. The analogue computer facility at Whetstone in Leicester also operated as a commercial computing centre and service and was available for use by staff from other divisions of English Electric and customers from industry. H. A. Darker, "Analogue computing: equipment and performance", *The English Electric Journal*, Vol. 18, no. 5, Sept-Oct 1963, pp. 31–39; D. Welbourne, *Analogue Computing Methods*, Pergamon, Oxford, 1965.

45. S. H. Lavington, *Early British Computers*.

46. The US computing centres were located in Princeton, NJ and Los Angeles, CA. EAI's Belgian centre contained three of EAI's PACE computers, two 48 and one 80 amplifier machines. B. Murphy, "This Analogue Centre Serves European and British Industry," *Automatic Data Processing*, March 1959, pp. 1–4.

47. Anon., "Analogue Computers", *The Computer Bulletin*, Vol. 3, Dec 1959, p. 76.

48. In the early 1960s English Electric and Solartron also established commercial computing centres and Fairey Aircraft offered a commercial computing service.

49. Anon., "PACE Computers", *The Computer Bulletin*, March 1963, p. 150.

50. *EAI PACE User Group Survey*, Typescript report, c. 1962; *Minutes of 2nd PACE User Group Meeting*, CEBG Headquarters, Friars House, London, 9 Oct 1961.

51. Guided weapons, analogue and hybrid computer activity were moved from Luton to BAC's new guided weapons facility at Stevenage. In 1977 nationalization of the aircraft industry under the Labour government led to a merger in which BAC and Hawker Siddeley Aviation became British Aerospace (BAe). At BAe Stevenage the use of analogue/hybrid computation, largely in missile/target simulation studies continued until the late 1980s. A. R. Adams, *Good Company: The Story of the Guided Weapons Division of British Aircraft Corporation*, British Aircraft Corporation, Stevenage, 1976; D. Edgerton, *England and the Aeroplane: An Essay on a Militant and Technological Nation*, Macmillan, London, 1991; C. Gardner, *British Aircraft Corporation: A History*, Batsford, London, 1981; K. Hayward, *The British Aircraft Industry*, Manchester University Press, Manchester, 1989.

52. For example, in the early-to-mid 1960s this was happening throughout the various divisions of English Electric where company analogue computer development was slowing down and coming to a halt. In the mid-1960s when a large computer system was required for its Nelson Research Laboratory, English Electric bought a Solartron 247 analogue computer and when new equipment was needed at English Electric's Nuclear Power Division at Whetstone (the home of the massive Saturn computer see above) it was purchased from EAL.

53. Elliott too was among the commercial manufacturers of analogue computer equipment who in the 1960s discontinued development and bought commercial systems. Elliott bought its first commercial system in 1963—a 120-operational-amplifier PACE 231-R system from EAL. It was primarily intended for development work on the new Vickers VC10. Anon., "PACE Computers", *The Computer Bulletin*, March 1963, p. 150.

54. Shorts developed a small, inexpensive analogue computer system as a response to the inclusion of analogue computing techniques in the curricula in universities and technical colleges. The basic model, known as the Shorts Educational Computer, sold for £900. It had 8 (expandable to 15) computing amplifiers and operated to an accuracy of about 1 per cent. Anon., "Analogue Computer: Educational," *Engineering*, 1 Oct 1961, p. 439. In the late 1950s Shorts also developed a helicopter simulator driven by a bank of analogue computers. It was used primarily to train helicopter pilots. Anon., "Simulating Helicopter Flight," *Engineering*, 18 Oct 1957, pp. 500–501.

55. Anon., "Analogue Computer", *Engineering*, Vol. 188, Oct 1959. p. 476.

56. H. G. Conway, "Makers of Analogue Computers," *Engineering*, 23 Aug 1963, p. 231.

57. D. S. Terrett, "Makers of Analogue Computers," *Engineering*, 30 Aug 1963, p. 260. The classification introduced by EAI seems to have been generally accepted. Even Solartron the only other British firm that could have contested the statement did not do so openly. Instead they described their machines as also belonging to the Class 1 category.

58. Computer Consultants, *The European Computer Users Handbook*, section 7.
59. RAL manufactured the all solid state Comcor Ci-5000, Ci-500, Ci-175 analogue/hybrid computers and the Ci-150 analogue computer. *RAL, Facilities and Services of Redifon-Astrodata Ltd.*, RAL, Littlehampton, 1966.
60. The AD4 computer systems remained in use at the RAE until 1982. Personal Interview, Jeff Baynham: Wellingborough, 30 April 1991.
61. The Royal Armament and Research Defence Establishment, Fort Halstead bought an AD/5 hybrid computing system from Applied Dynamics, for use in missile and projectile simulation. Ibid.
62. This table was complied from Solartron Trade publications and news items and articles in several journals; including, *Automation, Automation Progress, Computer Weekly, Electronic Engineering, Engineering, Journal of the Royal Aeronautical Society, Proc., IEE, Process Control and The Engineer*; also Personal Interview: Derek J. Cartwright, Hendon, Sussex, 1 May 1991.
63. Anon., "Analogue Computer: Capable of Expansion," *Engineering*, Vol. 191, 7 Sept 1962, pp. 307; Anon., "Characteristics of General Purpose Analog Computers", *Computers and Automation*, Vol. 13, June 1964. p. 78. The Solartron 247 range of computers cost between $28,000 and $350,000 (1964 prices) and consisted of the following models

Solartron 247 Series

Model	Description
Solartron 247	– basic unit expandable from 24 to 400 amplifiers, repetitive and iterative operation.
Solartron 247/36	– 60 amplifier system.
Solartron 247/72	– 150 amplifier system plus digital patch panel.
Solartron 247S	– 170 amplifier system all high precision, temperature controlled 0.01% linear and non-linear components, plus full digital logic options as standard.

64. In December 1963 Solartron announced the introduction of the Analogue Tutor Mark II for the educational market. Anon., "Improved Analogue Computer Trainer," *Engineering*, Vol. 196, 27 Dec 1963, p. 812; Z. Nenadal and B. Mirtes, *Analogue and Hybrid Computers*, SNTL, Prague 1962, English translation, Iliffe Books, London, 1968, p. 596–598.
65. Solartron Hybrid 7 Series 1 was not in itself a hybrid computer, but became one if used in conjunction with a Series 3, 6 or 7 machine.

Solartron Hybrid 7 Series

Model	Description
Solartron Hybrid 7 Series 1	– 24 amps, plus digital logic/patch panel.
Solartron Hybrid 7 Series 3	– 72 amps plus digital logic/patch panel and computer interface logic.
Solartron Hybrid 7 Series 6	– 144 amps plus digital logic/ patch panel and paper tape input for data and instructions and digital computer interface.
Solartron Hybrid 7 Series 7	– 144 amps plus all high precision components (0.01%) and full hybrid facilities for interfacing to digital computer.

66. This figure includes not only sales of computer equipment but also service and maintenance business. Personal Interview: Derek J. Cartwright, Hendon, Sussex, 1 May 1991.

67. In 1965 Solartron was taken over by the American Schlumberger Group of companies.

68. The total UK market in analogue and hybrid computers in 1968 was estimated to be between £1m and £2m annually. While the upper figure comes from estimates supplied by the individual firms of their annual business, the lower figure is a more realistic annual figure—except for those years when one or two especially large systems were sold. Personal Interview: Derek J. Cartwright, Hendon, Sussex, 1 May 1991.

69. Personal Interview, Derek J. Cartwright, Hendon, Sussex, 1 May 1991; Personal Interview: Keith Knock, Manchester, 4 March 1991; C. L. Boltz, "Business with a big future", *Computer Weekly*, Hybrids and Analogues Supplement, 11 Jan 1968, p. 1; D. Cartwright, "Growth of the technique", *Computer Weekly*, Feature Analogue/Hybrid Computing, 13 Aug 1970, pp. 9–12.

70. Anon., "Control Science at Cambridge", *Engineering*, 19 July 1968, p. 125.

71. At UMIST in 1966 a new chair in Control Engineering was created, and H. Rosenbrock moved from the Control Group at Cambridge University to take up the professorship. The following year the SRC awarded UMIST a grant to purchase two Redifon-Astrodata Ci 175 analogue computers which were originally to be interfaced to a Ferranti Argus 300 already in operation at UMIST. This scheme was soon abandoned because the Argus 300 was located too far (100 yards) from the analogue computers and the interface could not operate effectively over this distance. Instead, an Elliott 903 digital computer was purchased. This proved too small for the multivariable theory that Rosenbrock was developing, and in 1968 another application for SRC funds was made. In 1969 the SRC awarded the Control System Centre at UMIST a further £190,000 to purchase a DEC PDP-10 and a third Ci 175. The hybrid remained in operation until 1977. (The Control Science Centre also had a Solartron 247 and two small analogue computers.) Personal Interview: Dr. George Barney, UMIST, Manchester, 15 March 1991 Personal Interview: Professor Howard Rosenbrock, UMIST, Manchester, 25 April 1991; G. C. Barney and J. N. Hambury, "Developing the UMIST system", *Computer Weekly*, 13 Aug 1970, p. 10.

72. Anon., "Hybrids and Analogues" *Computer Weekly*, 11 Jan 1968, p. 1.

73. The Filton group was established in 1958. After the formation of the British Aircraft Corporation it became the Simulation Group of Engineering Computer Services, BAC, Filton. Simulation studies began with the investigation of the in-flight-handling characteristics of supersonic research aircraft and included hardware-in-the-loop applications. The group was also involved in the design and development of guided weapons—including the Bloodhound missile—as well as fighter and transport aircraft. In 1962 the Filton simulation group began work on the design phase of the supersonic passenger aircraft, Concorde. By 1972 the computing equipment for flight simulation included three EAI PACE 231-R and one TR48 analogue computers, totalling approximately 450 amplifiers. For engine performance and a variety of other studies an EAI 680 hybrid computer with 470 amplifiers was used. This equipment was not only used for the purpose of training flight crew; it was also used as a design tool that enabled the test pilots to evaluate the stability and handling characteristics of Concorde in complete safety. Modifications were then made to the design and reproduced in the simulator for re-evaluation. Anon., "Concorde Flight Simulation at Filton, British Aircraft Corporation, Filton Division, Engineering Computing Services, Typescript report, October 1972.

74. M. Brown, "Which System to Choose?", *Computer Weekly*, 11 Jan 1968, p. 5.

75. M. G. Brown, "An Introduction to Hybrid Computing", *Typescript report*, British Aircraft Corp., Stevenage, 1976; P. H. Enslow Jr. (Lt. Col.), "Hybrid Computer Technology in Europe: A State of the Art Survey", *ERO Technical Report No. ERO-1-74*, US Army European Research Office, London, 15 Feb, 1974.

76. COMPARISON OF DIGITAL AND ANALOGUE COMPUTATION

Digital computer	Digital simulation language	Digital computer (ratio to real time)	Analogue computer (ratio to real time)
IBM 7090	DAS	900	1
Ferranti Atlas	CIP	700	1
IBM 7090	FIFI	90	1
IBM 7090	MIMIC	60	1
IBM 7090	DSL 90	35	1
IBM 7090	CSMP	14	1

The test problem chosen was that of the equations describing a short circuit applied to an alternator. P. M. Pigott and R. J. Smale, "Hybrid Computer Study Using the EAI 8900 at EAI PCC", Minutes of the 25th Meeting: 26th July 1967, in *PUG Newsletter*, Vol. 1. no. 2, Sept 1967, p. 2. Comparing the speed of the CEGB's own IBM 360/75 and the EAI 8900 at EAI's Princeton Computing Centre in the USA, Pigott and Smale found that the hybrid system was approximately 150 times faster than the digital computer. They estimated that doing the same number of runs of the problem on the IBM 360/75 as were done on the EAI 8900 would have cost approximately £2 million in computer time. R. G. Blake, P. M. Pigott, R. J. Smale, R. K. W. Bowdler and M. J. Whitmarsh-Everises, "Hybrid Computer Simulation of Dungenesss 'B' Power Station", *Engineering Division Report, No. RD/C/R52*, CEGB, London, May 1967.

77. D. Millichamp, "Computers help power users", *Engineering*, 24 July, 1970, p. 88. The CEGB used the hybrid computer system for nearly eight years, after which time it was sold to Marconi Space and Defence Systems, Ltd., Stanmore, for approximately £100,000. Marconi Space and Defence Systems Ltd already had a EAI 8900 hybrid system obtained as part of an AUWE defence contract. Personal Interview: Keith Knock, Manchester, 4 March 1991.

CHAPTER 7

Electronic Analogue Computers and Engineering Culture

INTRODUCTION

In the previous chapters, developments in analogue and hybrid computation have been dealt with largely in the historical context of military and civil projects, user/developers firms and commercial manufacturers. Yet the military, political, economic and technical imperatives which helped to foster, fund, establish and sustain developments in analogue computation do not in themselves explain much of the appeal or importance of analogue computers in the context of engineering design.

In this chapter the emphasis is on the relationship between electronic analogue computing and the culture of engineering design. I argue that electronic analogue computers were shaped by a 19th and early 20th century American engineering pedagogy, characterised by the use of graphical methods, physical scale models and a preference for empirical trial-and-error and parameter-variation methods over theoretical and analytical approaches, and a preference for "doing" rather than "knowing". The analogue systems embodied these values and practices, and helped to enhance and extend the tradition.[1]

However, from the late 19th century the growing complexity and scale of technical systems increasingly challenged the efficacy of traditional engineering design practice and "brought the limits of empiricism to the fore".[2] Moreover, in the second half of the 20th century, in addition to the sheer complexity of problems, economic, safety and security considerations meant that many technical systems could not be designed by traditional trial-and-error and parameter-variation methods alone. Nor could they be designed entirely by analytical methods. Many phenomena were not well understood: there were too many unknowns and too many combinations of forces and circumstances to take into account. Engineering needed to bridge the gulf between scientific and engineering theory and the real-world technical systems that engineers were constructing.

The response of engineers was to construct a design tool that embodied many of the preferences, practices and values of their culture. This tool was the electronic analogue computer. By enrolling electronics to redefine model building techniques and enhance empirical design methods, engineers were able to push back the limits of empiricism, as well as the scientification of engineering. The electronic analogue computer went beyond its functions as an equation-solver and a system for modelling and simulating dynamic systems. From engineers' own accounts we find that electronic analogue computing enhanced engineering design by acting as a direct aid and guide to the thinking process. It was argued that as a result of their use and interaction with the analogue computer, engineers obtained not only a better "feel" for the problem, but also gained "insight" into where weaknesses in the design lay and the direction to follow in order to reduce them. In a sense the engineer and the electronic analogue computer held a conversation. Constructed by and in the image of engineering culture, the analogue computer was given a lab coat and invited into the laboratory. There, it helped engineers to build models of, test, think about and redesign the complex technical systems that engineers were constructing and trying to make work.

This chapter begins by reviewing the nature of American engineering culture in the late 19th and early 20th centuries. In particular, I address the predilection of engineers for empirical methods. This is followed by a discussion of the relations between electronic analogue computing and engineering.

ENGINEERING CULTURE IN THE USA IN THE 19TH AND 20TH CENTURY

In the latter half of the 19th century the engineering curriculum in American colleges was decidedly practical in its orientation. In the words of the historian Edwin Layton, curricula "placed as much or more emphasis on craft skills as upon scientific training." In support of his argument, Layton quotes a prominent mechanical engineer, Alexander L. Holley, who in 1875 asserted that college education was to train "not men of good general education, but artisans of good general education", and further that "the art must precede the science".[3] Though it was advocated from the 1890s onwards that more attention should be given to scientifically derived theory and mathematics, change was slow.[4] As Larry Owens points out, even at MIT in 1919:

> The curriculum stressed mechanical skill, work in the shop and laboratory, and mathematics and physics couched in the graphic idiom. The calculus was introduced to freshmen in a manner which emphasized graphical presentation and intuition over abstract rigor.[5]

The fact that engineering students were taught calculus at all is worth noting. As Layton has pointed out, until 1920 there was still considerable debate among engineering educators whether their students needed to learn calculus at all.[6] Indeed, the American engineering "art" of the late 19th century continued, in the 20th century, to stress the experimental approach and practice over theory.

Support for this view comes from Eda Kranakis' comparative study of 19th century American and French engineering cultures. This study shows that American engineers tended to avoid theory but devoted much of their attention to experimental and empirical research. Kranakis comments that it seemed as though:

> the attitude of 19th century American engineers towards theory, particularly mathematical theory, was often not just one of neglect but rather one of mistrust bordering on hostility.[7]

In her study, Kranakis quotes from a report on American technical education written by Omer Buyse, a Belgian who toured America at the turn of the century. Buyse wrote in 1908 that:

> The teaching of pure and applied science is imbued with the principles and methods of "rediscovery", practised in the laboratories and workshops. ... The student must wrest the secret of the phenomena and the laws which govern them directly from the experimental apparatus and equipment.[8]

As Kranakis points out, in the late 19th and early 20th centuries in the USA there was a widespread belief among engineers that, in order to uncover new laws or develop new theories, one only needed to compile masses of data in a coherent way.[9] American engineering characteristically worked out new designs by building a series of models and prototypes and from them "wrested" the "secrets" and "laws" of phenomena by direct experimentation. Yet we know that American engineers did use theory. However, as Kranakis has noted, engineers only considered theory important in so far as it was essential "to create 'successful' artifacts".[10] In the late 19th century engineers did not entirely disregard theory or theoretical knowledge, but they did pay more attention to empirical techniques and experimentation as a means of generating new designs and knowledge. For engineers, theory was no substitute for experience.[11]

To the discussion on the use of theory in engineering practice, Walter G. Vincenti adds: "It is of course true that engineers use theory whenever they find it feasible and advantageous to do so." However, it is often "... a matter of choice, not of necessity."[12] Vincenti argues that experimental parameter-variation techniques and their independence from physical theory gave engineers an alternative to theory. Vincenti carried out a study of the tests on aircraft propeller design undertaken between 1916 and 1926 by W. F. Durand and E. P. Lesley. In this study he analyzed the use of paramter-variation techniques in aeronautical engineering. He concluded that the strength and appeal of parameter variation is that "there is no essential relation between experimental parameter variation and physical theory".[13] Or in other words, that experimental parameter variation may be undertaken without recourse to physical theory and therefore could be used either in the absence of theory or to circumvent its use. Vincenti writes:

> Experimental parameter variation is used in technology (and only in technology) to produce the data needed to bypass the absence of a useful quantitative theory, that is to get on with the technical job when no useable theoretical knowledge is available.[14]

Moreover, Vincenti contends that even when analytical methods were applicable, engineers often chose to use experimental methods because:

> ... the experimental approach provides data relatively quickly, is comparatively free of the assumptions and simplification required by theory and can

help bring to light unforeseen problems; it has the disadvantage of requiring considerable experimental effort and therefore usually great cost.[15]

Indeed, in engineering, mathematics and theory were often thought of as either inappropriate or inadequate aids to design. Inappropriate, because the calculations required by analytical methods were often too laborious and too time-consuming, to be undertaken. Inadequate, because the pertinent theory had neither been established nor validated.

T. P. Hughes, in his study of the development and expansion of electric power distribution networks between 1880 and 1930, describes an example of a design problem where analytical methods actually were inadequate, and engineers resorted to empirical methods. Hughes writes about the early 20th century:

> A reverse salient of the ever-expanding, complex regional [electricity distribution] systems was the inability of engineers to precisely analyze and define them with equations that showed functional relationships. Efforts to write these equations resulted in complicated mathematical problems, the solution of which was tedious and time-consuming, if not impossible. Without a precise and clear understanding of the system, engineers had to rely on empirical, or cut-and-try, methods to improve systems performance.[16]

However, in the 20th century, traditional engineering practice, as well as analytical techniques, was increasingly challenged by the increasing complexity of technical systems.

Yet the growing complexity and scale of technological systems also demonstrated the limits of empirical design methods. In part this required engineering to rely more heavily on analytical theory and mathematics. This created a paradox for engineering design. On the one hand theoretical models detailed enough to reflect the complexity of the technological system were often too complex mathematically to be dealt with by existing methods of computation, and/or no useful theory existed. On the other hand the complexity of systems meant that they could not be designed entirely by trial-and-error and other empirical methods.

To this extent there was a gap between engineering theory/ mathematical techniques and existing empirical methods on the one

hand, and the complex real-world systems that engineers were designing on the other. Electronic analogue computers enabled engineers to
bridge this gap by augmenting empirical techniques, while also
enabling them to circumvent or overcome the shortcomings of theory.

Returning to the issue of engineering education, we find that in the
USA between the end of World War II and 1960, curriculum shifted
more rapidly and decisively in the direction of "engineering science"
than at any other time. In his study of American engineering college
system between 1900 and 1960, Bruce Seely notes:

> By 1960, ... engineering education looked very different. The emphasis on
> rule of thumb learned through practical experience had given way to an
> education stressing scientifically derived theory expressed in the language
> of mathematics.[17]

And yet, as the current analysis of the history of the electronic analogue computers shows, engineering continued not only to rely on,
but endorse experimental approaches and the graphical idiom, and
continued to value "doing" over "knowing".

THE ELECTRONIC ANALOGUE COMPUTER IN ENGINEERING DESIGN

One of the first engineers to explicitly enrol electronic analogue
devices as empirical tools in the design and analysis of technical
systems was George A. Philbrick, (see chapter 3). Working at the
Foxboro Instrument Co. in the late 1930s, Philbrick recognised that
the increasingly complex and mathematical orientation of control
theory created problems for engineers involved in the design of industrial process controllers. Philbrick thus designed a special-purpose
electronic analogue computer, which he called the Automatic Control
Analyzer (ACA). Built between 1938 and 1940 the main purpose of
the analyzer was to enable engineers to solve problems in the design
of process controllers empirically, and thus to gain a better understanding of their operation. In Philbrick's own words:

> The user [of the ACA] could select—by means of a moment's manipula-
> tion—the best type of controller and the best adjustments for that con-
> troller to apply to a given process. ... Coupled with hard-boiled
> experimental evidence from the field, this machine will replace years of
> arduous calculation (and dangerous guesswork) in the development of
> automatic controllers ...[18]

Yet Philbrick saw the ACA as having considerably more potential. In the education of engineers in the ways of process control, the ACA also afforded:

> ... a Royal Road to knowledge. While there is no substitute for experience, it is held that the kind of process control information which can be easily acquired from the Analyzer goes a long way in this direction. Demonstration of the dynamic characteristics of processes, hitherto the work of mathematicians, can be made by anyone who is able to follow a flow diagram. The effect of varying amounts of reset or control, for example, is shown in a lively graphic form which is better than any amount of wordy description. Every phenomenon of Automatic Control may be pictured at will, and with no more effort than is needed to turn knobs.[19]

The solutions were displayed by means of a single five-inch oscilloscope. Here solutions appeared in a "lively graphic form" that engineers could interpret without recourse to "wordy descriptions" or indeed to mathematicians. Yet Philbrick's analyzer was not unique in presenting results in a graphical form: the plotting tables of the mechanical differential analyzers had been doing this for some time. What Philbrick demonstrated with the ACA was the advantage of combining the graphical idiom with the ability to rapidly alter parameters and even more rapidly recompute the effects, in other words how "every phenomenon ... may be pictured at will, and with no more effort than is needed to turn knobs."

With the ACA, Philbrick demonstrated how active electronic components could be used to build models, and how high-speed computation facilitated trial-and-error and parameter-variation methods. Moreover, he was indicating how high-speed empiricism could be substituted for theory and mathematical rigour in engineering design. Briefly stated, the ACA helped engineers deal with the increasing complexity of control theory by minimising the mathematical and emphasising the experimental and visual content of the design process.

Philbrick's interest in electronic models and high-speed computation continued after the war, and he set up his own company to manufacture modular electronic analogue computing devices and components. In 1952 Philbrick's firm published the first issue of its own "aperiodical" journal, called the *Lightning Empiricist*. The *Lightning Empiricist* was co-authored by Philbrick and Henry M. Paynter, a Professor of Mechanical Engineering at MIT and a business partner of Philbrick's. Their journal was aimed at:

> devotees of high-speed analogue computation, those enthusiasts for the new doctrine of Lightning Empiricism.[20]

Philbrick and Paynter were indeed devotees of the new doctrine, and they had much of the religious zealots about them. Philbrick and others had already converted feedback, operational amplifiers and oscilloscopes to the cause of lightning empiricism, now they were hoping to enlist others in the ways of electronic models. Their publications were couched in terms that would have had little appeal to those with a scientific, progressive view of engineering. Instead, they echoed much of Philbrick's prewar enthusiasm for the ACA, model-building and empiricism and encouraged a view of analogue computing as more art than science:

> If Mathematics can be called the "Queen of the Sciences", then truly Analogy is her consort. When logic is enlivened, existence theorems become self-evident. By introducing *amplifiers* into physical structures to supply activation energies while enforcing signal flow causality, unlimited realms of abstraction may be physically realized. This does not mean, however, that Fictions may be made Truths; indeed learning to flatter rather than antagonize Nature is part and parcel of the Analog Art. ... using active circuits, one may construct an *Electronic Analog Computor*, a device using voltages to represent all variables. ... If relations and parameters are known, computation is a straightforward problem of solving algebraic and differential equations. However, analog computors also permit studies of systems where relations describing performance are not clearly known in advance. In this case portions of the system may be *simulated* by active or passive networks constructed experimentally to give a behaviour approximating known or desired response.
> ... Howsoever they arise, all computor representations of real or abstract systems are physical *Models*, but of extraordinary flexibility.[21]

Philbrick and Paynter, though perhaps the most colourful, were by no means the only devotees of high-speed analogue computing. As we have seen, in the late 1940s electronic analogue computing underwent a period of rapid transformation from small experimental systems consisting of a few operational amplifiers to commercial general-purpose electronic analogue computer systems and large-scale installations.

The chief motivation for the systemisation of operational amplifiers into general-purpose electronic analogue computers was the growing complexity of modern missiles and high-speed aircraft, coupled with the desire to reconstruct the conditions of the field trial in the engineering laboratory where complex dynamic systems could be designed, modeled and simulated hundreds of times, economically and safely.

Much of the increased complexity in post-World War II aeronautical design arose out of the higher flight speeds of military aircraft and missiles, and the development of closed-loop control systems for missile guidance and auto-pilots. In the development of aircraft and missiles, higher flight speed emphasised non-linear effects. Previously, non-linearities were dealt with by making assumptions and simplifications and using "linearization" techniques. In aircraft, the instabilities that resulted from residual non-linear effects could be taken care of by the pilot. However, in the design and development of increasingly complex aircraft and missile systems, the inadequacy of existing techniques was becoming more and more apparent. Reviewing the situation in 1946, Professor A. C. Hall, the Director of the Dynamic Analysis and Control Laboratory at MIT, commented:

> The complexity is such that general solutions are not known and few specific cases have been solved without drastic assumptions.[22]

The development of guided missile systems generated a number of unique and particularly vexing problems. Control systems had to be designed to operate under, and compensate for, a great variety of atmospheric conditions. Electrical "noise" in the control systems and the variability of components were also among the random events that affected the performance of a missile (measured largely in terms of its ability to hit its target). In order to optimise the missile system, a large volume of data was required for statistical analysis. To compile such data by trial-and-error methods involving hundreds or even thousands of flight trials would have been enormously expensive and quite impractical. From the mid-1940s through the 1950s the problems faced by missile designers remained largely the same. Stanley Fifer, the former Head of Project Cyclone, summarised the problems faced by missile designers as follows:

> [In missile systems design] ... The existing non-linearities render thorough mathematical analysis quite fruitless, and resort to intuition alone quite dangerous, while the complexities of the problem make solution by numerical methods impractical. Moreover, the number of solutions required to evaluate the overall system performance and to determine the optimum values of parameters may be in the order of hundreds or even thousands.[23]

This was the design context from which general-purpose electronic analogue computers emerged (see chapter 4). An indication of the interrelatedness of postwar electronic analogue computing and the values, practices and preferences of engineering culture can be seen in the following quotation. This is taken from the first public release of information on the Reeves REAC and was published in 1948:

> Problems can usually be stated mathematically, but obtaining numerical solutions is often impractical because of the enormous amount of computation that is required. In the field of electronics, for example, the relationship between the current and circuit elements of a tuned plate oscillator is ... [the differential equation is then given] ... Most of us find it easier to build such an oscillator and use cut-and-try methods in selecting its components, than to solve this second-order differential equation. If the equipment is [the] goal this approach is satisfactory, but when a fundamental design principle is being studied, the exact solution is required.
>
> The REAC ... can solve such problems, and more complex ones, in a relatively short time. Once the problem has been set into the REAC, the effect upon oscillation of different values of the circuit parameters can be obtained simply by turning a dial; solving the equation is thus simpler and neater than changing components on a breadboard.
>
> What is actually done when an equation is set into the REAC is to simulate the system to which the equation applies. For example, using aerodynamic relationships expressed in terms of ordinary, initial valued, linear and nonlinear simultaneous differential equations, the REAC can be set so that it simulates the flight of an aircraft. The effect of contemplated design changes, or the design that will give optimum performance, can be determined without actually building an aircraft or [physical] model.[24]

Here we see that there was a clear preference for "cut-and-try" (more commonly referred to as trial-and-error) design methods rather than solving differential equations—even though both methods may have been equally laborious. The appeal of the REAC, and postwar electronic analogue computing in general, was that it enabled engineers to circumvent the actual or perceived impracticality of obtaining numerical solutions. They automated mathematics, whereby any number of solutions could be obtained by "turning a dial". Moreover, after some redefinition, much of the traditional model-building and "cut-and-try" engineering design methods could be retained.

In 1961 Fifer summarised the contribution of the electronic analogue computer to engineering design as follows:

> The analogue computer plays many roles: ... Perhaps its most important contribution has been to replace expensive, time-consuming, cut-and-dried design methods by economical, rapid and systematic laboratory exploration procedures.[25]

In a general sense this was the postwar response of engineers to the complexity of technical systems and the limits of empiricism. Electronic devices developed during the war to solve one specific mathematical problem—how to guide and control an anti-aircraft gun—were redefined as mathematical operators which were in turn developed into a generalised system for computing by electronic analogy. The electronic analogue computer enabled engineers to bring the missile and the aircraft into the laboratory where, through "systematic laboratory exploration procedures", "masses of data" could be compiled and their "secrets" and "laws" could be "wrested" from them.[26]

Yet in the words of Granino and Theresa Korn, who in 1952 authored the first comprehensive text on electronic analogue computation:

> The really important contribution of DC analogue computers to modern research and development techniques goes beyond mere numerical computations. In many applications, the analog approach functions as a direct aid to a research worker's or engineer's thinking process: an analogue-computer setup serves as a *model* which helps to bridge the gap between mathematical symbolism and physical reality.[27]

Indeed, in the hands and minds of the design engineer, the analogue computer was more than a high-speed equation solver.

Analogue computers did more than provide a new empirical design tool. They resonated with and fostered an engineering design culture characterised by tacit and non-verbal knowledge, rules of thumb and a reliance on having an intuitive "feel" for problem-solving.

In his book, *Engineering and the Mind's Eye*, Eugene Ferguson has argued:

> Until the 1960s, a student in an American engineering school was expected by his teachers to use his mind's eye to examine things that engineers had designed—look at them, listen to them, walk around them, and thus to develop an intuitive "feel" for the way the material world works (and sometimes doesn't work).[28]

Indeed, historians of technology have identified intuition, tacit knowledge and visualisation as central components of traditional engineering culture. Yet from the late 1940s to the 1970s we find that engineers themselves described and discussed how electronic analogue computing enabled them to walk around and gain "insight" into problems. They discussed the nature of their relationship with the analogue computer and stressed that analogue computing techniques helped engineers to develop an intuitive "feel" for problem-solving. Here the intimacy of the relationship was central. The following quotation addresses this issue:

> The use of an analogue computer to facilitate engineering design synthesis has certain advantages over numerical and analytical methods, the machine method of solution being intuitively straightforward and logical. Once the analogue has been established in the engineer's mind by the initial presentation of the equations of a dynamic system to the machine, changes in system design can be accomplished by reconnecting machine elements with little if any reference to the equations which by implication are being changed at the same time. The analogue computer becomes the physical system in the operator's mind, so that gear ratios can be changed, new pickups installed, linkages redesigned, altitude and airspeed changed, wings relocated, etc. merely by reconnecting electrical leads and changing coefficient potentiometer settings. Not only does this permit a quicker design process, but the intimacy of the engineer with the problem promotes improved design philosophy and insures against the gross errors that sometimes result when abstract analysis goes astray.[29]

By the mid-1950s there was a growing consensus that a significant advantage of electronic analogue computing *vis-à-vis* digital computing was its intimate and interactive nature (in chapter 8 the analogue versus digital question is discussed at greater length). In the study of dynamic systems by simulation, this interactive relationship was seen as essential, and was likened to a conversation:

> An element which is essential to the process of engineering design by simulation is the "conversational" nature of the interaction between the analyst and his simulation model.[30]

Yet for the conversation to be lucid, several factors were important: the speed of the computer (how long it took to answer the engineer), the ease with which parameters could be varied and the model modified, and the graphical display of results which enabled the results of changes to be rapidly interpreted. An important factor was the accessibility of electronic analogue computer systems (see the discussion on "open" and "closed-shop" computing in chapter 8). Direct and easy access to the computer meant that through experimentation the engineer could rapidly gain "hands-on" experience of the dynamic system being studied.

From the perspective of engineering practice, one of the most important aspects of the close, interactive nature of electronic analogue computing was that it incorporated and facilitated the contingent and stochastic character of the design process. Electronic analogue computers allowed new ideas to be tested at once, and were described as enabling a design engineer to "think as he goes", and to "give his hunch a chance". Yet perhaps the most significant contribution to engineering design was that the conversation between engineer and analogue computer was seen as enhancing the engineer's understanding of complicated dynamic systems. Much of the knowledge that engineers gained was, it seems, tacit knowledge. In the following two quotations the appeal of analogue computation is couched in terms of the stochastic and contingent character of design, and the tacit and intuitive aspects of engineering knowledge. The first quotation is from an article on the application of electronic analogue computers to engineering problems and was published in 1956:

For the engineer, probably the most important appeal is that the analog computer provides *a ready means for developing insight into the operation of complicated systems*: any part of a system can be changed easily and the effect observed. Closely allied with this factor is the ability of the engineer to *"give his hunch a chance"*.[31]

Almost ten years later, in an introductory text on electronic analogue computers, these aspects of electronic analogue computing and engineering design were still central to their appeal:

Probably the chief advantage in analog computation is that the operator retains a *"feel" for his problem* ... Changes in settings of computer components thus become meaningful in terms of the real process, and the result of these changes can be interpreted immediately in the same terms. The [electronic analogue computer] operator can *"think as he goes,"* and if interesting side-avenues open up, these can be immediately explored.[32]

Thus, from the late 1940s to the 1970s, electronic analogue computers enhanced engineering design in several ways. They permitted dynamic systems to be reconstructed in the laboratory, where they could be designed and tested empirically, and thus helped engineers to derive solutions to problems within the time scale and economic constraints of the "real-world" engineering environment. They helped engineers to bridge the gap between abstract mathematics and physical reality, and thus to overcome the increasing complexity of technical systems. The key characteristics that kept the conversation between the engineer and the analogue running, were computing speed, the ease with which parameters could be changed and modifications made to the model, and the corresponding responses observed.[33]

To paraphrase from Philbrick and Paynter, through interactive, model-oriented, empirical design, engineers learned to "flatter rather than antagonize Nature". Indeed, this was "part and parcel of the 'Analog Art'" which the postwar "Lightning Empiricists" practised.[34] What we see is that engineers enrolled electronics to enhance their own traditional "Art", and in doing so were able to push back the limits of empiricism and the scientification of engineering. Yet there was also a limit to the extent that electronic analogue computers, could continue to compensate for the increasing complexity of technical systems.

AUTOMATING THE ANALOGUE ART: COMPLEXITY OVERWHELMS INTERACTIVE ANALOGUE COMPUTING

In the USA, until the early 1960s the real-time simulation studies of dynamic systems, such as missiles, aircraft, space vehicles, as well as nuclear and chemical plants, was almost entirely performed with electronic analogue computers. During the development of these technical systems, much was learned that added, both to the complexity of subsequent systems, and to the models used to describe them. The continuing growth in system complexity created a paradox for electronic analogue computation. Electronic analogue computers were well suited, because of their parallel architecture, to very high computing speeds and to the real-time simulation and study of large, complex dynamic systems. But the physical size of analogue computer installations had to grow almost linearly with the increase in size and/or complexity of the design problem. Consequently, the benefits of interactive analogue computing began to be undermined by the sheer size and complexity of the electronic models needed to represent such systems.

Computer setups grew from between 20 to 30 operational amplifiers in the late 1940s to over 300 in the late 1950s, and the number of interconnections increased from tens to hundreds, and even thousands. As a result, the process of setting up the electronic analogue computer, and checking it, became increasingly complicated and time-consuming. The size and complexity of the analogue computer and the electronic model meant that it became increasingly difficult for the design engineer to gain "insight", or get a "feel" for the problem.

The engineering response to the limits of electronic analogue computing, and the gradual deterioration of their "conversation", was to enrol the digital computer as an intermediary. Digital computer programs were written to convert the equations of mathematical models into a list of interconnections, parameter settings and test values that could be used to check the manual computer setup.[35] Yet this use of digital computers did not fundamentally alter the interaction between the engineer and the analogue computer. Once the analogue computer had been set up, the study was carried out entirely on the analogue computer as normal.

FIGURE 7.1 PREPARATION OF THE PATCH PANELS FOR A LARGE ANALOGUE COMPUTER SET-UP

However, with the development of hybrid computers, the "conversation" between the engineer and the computer model, was fundamentally changed.[36] The electronic analogue computer was no longer operated directly by the engineer, but took its instructions from the digital computer, which systematically performed a predefined set of experiments, varied parameters, and performed statistical analysis and system optimisation.[37] Though many welcomed the shift towards digital computer control, there were also those who recognised the dilemma of automation through hybridisation:

> Since the main field of analogue computation is in problems where further insight into the problem is of importance, much of this would be lost if the preparation of the associated analogue circuits were not undertaken by the human analogue computer programmer, who also has an interest in the results.[38]

Yet, though this remained true for the study of small systems, for large, complex models the benefits of digital computer control of the analogue computer were generally seen as far outweighing any possible loss of insight. As a tool for engineering design, the electronic analogue computer was overwhelmed by the increasing size and complexity of modern technological systems.

CONCLUSION

The electronic analogue computer was constructed in the image of an empirical design tradition in engineering, to help engineers deal with the increasing complexity of dynamic systems. Yet though DC operational amplifiers did no more than operationalise mathematics, the electronic analogue computer did much more than compute: it resonated with and sustained an engineering pedagogy that emphasised empiricism and graphical methods over mathematical rigour. Insight and intuition were central to the "Analog Art", and through electronic analogue computation its Artisans frustrated the scientification of engineering. But the Art was automated, and without the interaction between analogue and engineer, the electronic analogue computer's role as a direct aid and guide to the thinking process largely came to an end. Built by engineers to serve engineering design, it is perhaps not surprising that engineers who had became accustomed to the intimacy and immediacy of electronic analogue computing, were willing to contend the relative merits of electronic analogue and digital computing. In the following chapter we see that the debate was a lively one, and at times bordered on the hostile. Engineers seemed not only to be defending their choice of analogue over digital computing equipment, but also the established traditions of the engineering culture to which they belonged.

NOTES

1. E. T. Layton, Jr., "Technology as Knowledge", *Technology and Culture*, Vol. 15, no. 1, 1974, pp. 31–41, pp. 33–34.

2. E. Kranakis, "Social Determinants of Engineering Practice: A Comparative View of France and America in the Nineteenth Century", *Social Studies of Science*, Vol. 19, 1989, pp. 5–70, p. 53.

3. E. T. Layton, Jr., *The Revolt of the Engineers: Social Responsibility and the American Engineering Profession*, Johns Hopkins Univ. Press, Baltimore, 1986, p. 4.

4. B. Seely, "Research, Engineering, and Science in American Engineering Colleges: 1900–1960", *Technology and Culture*, Vol. 34, no. 2, 1993, pp. 344–386, p. 379; E. T. Layton, Jr., "Mirror Image Twins: The Communities of Science and Technology in Nineteenth Century America", *Technology and Culture*, Vol. 12, no. 4, 1974, pp. 562–580.

5. L. Owens, "Vannevar Bush and the Differential Analyzer: The Text and Context of an Early Computer", *Technology and Culture*, Vol. 27, no. 1, 1986, pp. 63–95, pp. 93–94.
6. E. T. Layton, Jr., *The Revolt of the Engineers*, p. 4.
7. E. Kranakis, "Social Determinants of Engineering Practice: A Comparative View of France and America in the Nineteenth Century", p. 42.
8. Ibid., p. 20.
9. Ibid., p. 44.
10. Ibid., p. 42.
11. R. Kline, "Science and Engineering Theory in the Invention and Development of the Induction Motor, 1880–1900", *Technology and Culture*, Vol. 28, no. 2, 1987, pp. 283–313; E. T. Layton, Jr., "Scientific Technology, 1845–1900: The Hydraulic Turbine and the Origins of American Industrial Research", *Technology and Culture*, Vol. 20, no. 1, 1979, pp. 64–89.
12. W. G. Vincenti, "The Air-Propeller Test of W. F. Durand and E. P. Lesley: A Case Study in Technological Methodology," *Technology and Culture*, Vol. 20, no. 4, 1979, pp. 712–751, p. 748. This is one of three studies of engineering practice by Vincenti, see also W. G. Vincenti, "Control-Volume Analysis: A Difference in Thinking Between Engineering and Physics," *Technology and Culture*, Vol. 23, no. 1, 1982, pp. 145–174; W. G. Vincenti, "The Davis Wing and the Problems of Airfoil Design: Uncertainty and Growth in Engineering Knowledge," *Technology and Culture*, Vol. 27, no. 4, 1986, pp. 717–758. Or see, W. G. Vincenti, *What Engineers Know and How They Know It: Analytical Studies from Aeronautical History*, Johns Hopkins University Press, Baltimore, 1990. Also see E. W. Constant, "Scientific Theory and Technological Testability: Science, Dynamometers, and Water Turbines in the 19th Century," *Technology and Culture*, Vol. 24, no. 1, 1983, pp. 183–198. In Constant's analysis of the role of testing in the development of water turbine technology in the 19th century, he concludes that "technological testing for the first time permitted practitioners to know which designs and which modifications represented real progress and precisely how closely they were approaching the theoretical ideal." p. 196.
13. W. G. Vincenti, "The Air-Propeller Test of W. F. Durand and E. P. Lesley: A Case Study in Technological Methodology", p. 748.
14. Ibid. (1979), p. 743.
15. Ibid. (1979), p. 744.
16. T. P. Hughes, *Networks of Power: Electrification in Western Society, 1880–1930*, Johns Hopkins University Press, Baltimore, 1983, p. 376.
17. B. Seely, "Research, Engineering, and Science in American Engineering Colleges: 1900–1960," p. 345.
18. G. A. Philbrick, "The Philbrick Automatic Control Analyzer: Program for Development and Exploitation", *Hand written memorandum, the Foxboro Co., RD Project 10530*, 1 Feb 1940, pp. 1–3, p. 1. Archives of the National Museum of American History, Washington, DC, Box: Philbrick.
19. G. A. Philbrick, "The Philbrick Automatic Control Analyzer", p. 2.
20. G. A. Philbrick and H. M. Paynter, *The Lightning Empiricist*, George A. Philbrick Researches Inc, Boston, Mass., No. 1, June 1952, p. 1.
21. G. A. Philbrick and H. M. Paynter, *Applications Manual for Philbrick Octal Plug-in Computing Amplifiers*, George A. Philbrick Researches Inc, Boston, Mass., 1956, p. 1.
22. A. C. Hall, "A Generalized Analogue Computer for Flight Simulation", *Trans., AIEE*, Vol. 69, 1950, pp. 308–312, p. 308.
23. S. Fifer, *Analogue Computation: Theory, Techniques and Applications*, 4 Vols., McGraw-Hill, New York, 1961, Vol. 4, p. 1087.
24. S. Frost, "Compact Analog Computer", *Electronics*, July 1948, pp. 116–120, p. 116.

25. S. Fifer, *Analogue Computation: Theory, Techniques and Applications*, Vol. 1, p. 1.

26. E. Kranakis, "Social Determinants of Engineering Practice", p. 20.

27. Korn and Korn continued: Simple patch-cord connections and potentiometer settings on a d-c analog computer enable the system designer to create a scale model which permits convenient investigation of the performance and interaction of components and system. New ideas can be tested at once: blocks of computing elements represent system components and are readily replaced or assembled into different system designs. Design parameters may be changed instantly by corresponding resistance changes, and high computing speeds permit optimization of system performance by successive parameter adjustment. G. A. Korn and T. M. Korn, *Electronic Analog Computers*, 2nd ed., McGraw-Hill, New York, 1956, p. 6.

28. E. S. Ferguson, *Engineering and the Mind's Eye*, MIT Press, Cambridge, Mass., 1992, p. 169; see also E. S. Ferguson, "The Mind's Eye: Nonverbal Thought in Technology", *Science*, Vol. 197, 1977, pp. 827–836.

29. Anon., "The Boeing Electronic Analog Computer: The use of analog equipment in the study of dynamic systems", *Internal typescript report*, Physical Research Unit, Boeing Airplane Co., Seattle, c. 1950, p. 3.

30. T. D. Truitt, "A Discussion of the EAI Approach to Hybrid Computation", *Simulation*, Oct 1965, pp. 248–257, p. 257.

31. P. J. Hermann, K. H. Starks and J. A. Rudolf "Basic Applications of Analog Computers", *Instruments & Automation*, Vol. 29, March 1956, pp. 464–69, p. 469.

32. J. E. Stice and B. S. Swanson, *Electronic Analog Computer Primer*, Blaisdell Publishing Co., New York, 1965, p. 3–4.

33. D. Rubinfien, "What Computers Promise", *Control Engineering*, Sept 1954, pp. 64–67.

34. G. A. Philbrick and H. M. Paynter, *Applications Manual for Philbrick Octal Plug-in Computing Amplifiers*, p. 1.

35. A. Debroux, C. H. Green and H. D'Hoop, "Apache—A Breakthrough in Analogue Computing", *Trans., IRE, Electronic Computers*, EC-11, Oct 1962, pp. 699–706.

36. G. A. Korn,"The Impact of the Hybrid Analog-Digital Techniques on the Analog Computer Art", *Proc. IRE*, Vol. 50, no. 5, 1962, pp. 1077–1086.

37. T. D. Truitt, "Hybrid Computation ... What is it? ... Who Needs it? ...", *Proc., Spring Joint Computer Conference*, 1964, pp. 249–269, p. 267.

38. W. G. Proctor and M. F. Mitchell, "The PACE Scaling Routine for Mercury", *The Computer Journal*, Vol. 5, 1962–63, pp. 24–27, p. 27.

CHAPTER 8

Negotiating a Place for Electronic Analogue Computers: the Analogue versus Digital Debate

INTRODUCTION

Parallel developments in electronic analogue and digital computers from the late 1940s to the early 1970s often led to comparisons of their relative merits being made. In the USA and Britain the ensuing debate was remarkably similar both in terms of the arguments that were marshalled and the manner in which the discussion changed over time. As already discussed in chapter 1, electronic analogue computing has received little attention from computer historians. The

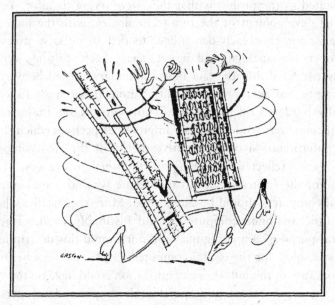

FIGURE 8.1   ANALOGUE VERSUS DIGITAL TECHNOLOGY![1]

contemporary debate has also been neglected. One is left with the impression that the analogue computer had few serious adherents. The presentist perspective from which existing accounts are written, implicitly and prematurely relegates analogue computing to the status of a "failed technology".[2]

Yet a number of factors, including the debate and the decisions of firms, individuals and universities to commit time, money and effort to the development of electronic analogue computing, suggests that there is a lacuna between contemporary attitudes and current historical retrospective. Indeed, contemporary attitudes indicate that the superiority of digital computing was equivocal, and the fact that electronic analogue computing would be superseded by digital was not accepted as a foregone conclusion. For more than twenty five years electronic analogue and digital computers were competing technologies.

Analogue versus digital comparisons were commonplace in internal reports, engineering text books, journals and the proceedings of the professional engineering societies. At one level, the debate can be seen as an attempt by participants to inform their fellow engineers and scientists as to which computing technology and technique was best suited to the problems that they were trying to solve. Yet the parallel development of the two technologies, with the increasing specialisation that each demanded, tended to polarise users and developers into analogue and digital "men". Such positions were not necessarily fixed, but the fact that a boundary existed between the two camps was acknowledged, and relations between the two occasionally bordered on the hostile. Those writing on analogue and digital computers tended to have different perspectives which shaped their interpretation of the relative merits of the two computing methods and reflected their own pre-established preferences.

One of the earliest issues in the debate was how analogue and digital computers should be categorised. More specifically, whether electronic analogue computers could justifiably be described as general-purpose computing machines, or should this description be reserved solely for the digital computers. This issue became important because of the influence categorisation could have on the status of the groups of analogue and digital computer developers and users.

However, the dominant issues in the debate were the comparisons of analogue and digital computing speed and the accuracy and precision of data and computation. These terms, while appearing to be overtly technical and quantitative, were in fact negotiated and subject to redefinition and reinterpretation. There was also an explicitly qualitative aspect of the debate, which stressed the interactive nature of electronic analogue computing, its one-to-one correspondence with the real-world systems under study, and the "feel" and "insight" which analogue computing gave users into the problems that they were investigating. In chapter 7, this aspect has been discussed in the context of the electronic analogue computer's relations to engineering culture.

At another level, the debate can be viewed as a manifestation of attempts by proponents of the two technologies to establish and retain "application domains" for their own technology. Speed and accuracy/precision were redefined, negotiated and made contingent to sustain these "application domains". Here I use the term "application domain" to refer to the perceived, and actual, landscape of problems that either computer technology was capable of solving. Application domains were fluid and overlapped, and advocates of each technology sought to maximise the perceived scope of their domain through comparisons with the competitive technology.

In chapter 7 I argued that postwar electronic analogue computers embodied and enhanced a pre-World War II American engineering pedagogy. Walter G. Vincenti has argued that an important aspect of engineering culture is its willingness:

> to forego generality and precision (up to whatever limits of accuracy are set by the requirements of the application) and to tolerate a considerable phenomenological component...[3]

This, Vincenti argues, differentiates engineering from scientific culture in that scientists value:

> generality and precision for their own sake and consider[s] that the theory or model be as close to first principles as possible.[4]

Similarly Edwin T. Layton Jr has argued:

> Basic science aims at knowing, it seeks generality and exactitude, even at the price of a good deal of idealization. [But] engineering science serves

the needs of practice, even when this involves loss of generality and acceptance of approximate solutions.[5]

In this chapter we see that when the relative merits of analogue computing were debated, neither generality, speed nor accuracy/precision
were necessarily valued for their own sake. The electronic analogue
computer community argued for a contingent view of speed and
accuracy/precision.[6] It forwent "generality and exactitude" and,
while serving "the needs of practice", accepted "approximate solutions". In doing so, the postwar electronic analogue computer community demonstrated that fundamentally it belonged to an
engineering culture rather than a scientific one.

This chapter describes three issues in the analogue/digital computer
debate: generality, speed and accuracy/precision. I analyze the debate
in terms of the actions of the analogue computer community in the
context of the competition for status and application domain. I also
examine how the debate changed its focus with the development of
hybrid computing systems. Here we find that the debate approached
closure both through the hybridisation of analogue and digital computers and through the displacement of analogue and hybrid computers by all-digital ones.

GENERAL-PURPOSE VERSUS SPECIAL-PURPOSE

The concurrent development of electronic analogue and digital computers that followed the end of World War II had by the early 1950s
led to commercial computing systems of both kinds. Yet, by the mid-
1950s, considerable dissatisfaction had already built up among proponents of electronic analogue computing. This arose, in part, from
what they perceived as a general bias, in the technical press, towards
all forms of digital computing.[7] There was also a feeling among the
analogue computing community that proponents of digital computing viewed the electronic analogue computer as outdated and as an
anachronism with little, if any, real future. The following quotation
from an article published in 1954 sums up quite succinctly the
growing feelings of indignation on the part of the analogue computer
community:

> So much has been written recently about the truly wonderful achieve-
> ments in the field of digital computing that there has been a tendency to

forget about analogue computers and to overlook the progress they have been making. Indeed there are those who would say that the analogue computer is outdated and dispute the need to improve it further. It is hoped ... to show that the need still exists, and further that the digital computer as it stands, in spite of its undoubted superiority in many cases, still has a long way to go before it can surpass the analogue device in all applications.[8]

Almost a decade later, when John McLeod, founding editor of the *Simulation Council Newsletter*, asked Lee Cahn of the Beckman Instrument Company to comment on the state of analogue computing, Cahn replied:

Perhaps the most significant thing to me is that analogue computing is alive at all, and thriving ... You may recall ... that ten years ago the digital boys were all saying that analogue computing was a crude anachronism, which would shortly be swept away by the manifest superiority of digits. Evidently this has not happened ... it does not appear that any function they ever had has been taken away.[9]

The analogue computer community was not only sceptical of reports of digital superiority, and analogue outdatedness, but were keen to refute—even ridicule—the claims of the digital "men". Though commercial competition helped fuel the analogue/digital debate, with the benefit of hindsight we can conclude that status was one of the underlying issues in the debate. At issue was the relative status of developers and users of electronic analogue computer equipment *vis-à-vis* the digital computer community of developers and users. One way in which this came to the fore was in the discussion over whether analogue computers could be categorised as general-purpose computers as opposed to special-purpose computing devices.

Post-World War II electronic digital computers were viewed by many as the only truly general-purpose computing technology and technique.[10] By implication, all other computing systems were special-purpose and of only limited use. Yet neither the users nor the manufacturers of electronic analogue computers welcomed the prospect of being relegated to the margins where obscure, specialised computing instruments lay. Rather than accept a subordinate view of electronic analogue computing, one way in which the analogue computer community responded was to attempt to redefine digital computing, so that it appeared as a mere variant on the more general category of computing by analogy:

> At first glance it does indeed seem that two separate technologies are in competition for supremacy. A closer examination, however, reveals that the digital approach is itself one specific class of analogue instrumentation. It deals with a group of problems that can be classified as number-handling problems ... One approach to the solution of these is to set up an analogue of a number system and to instrument the rules of arithmetic for the manipulation of numbers ... Thus, rather than constituting a flaw or discrepancy in the validity of the concept of analog operators, the high state of development of digital computers and machines is a demonstration of the scope of the more general category.[11]

In the climate of competition between analogue and digital computers, electronic analogue computer manufacturers and users challenged the images of the analogue and digital computers. They preferred their computers to be seen as versatile, flexible and fast. The view that the analogue computer community promoted was that the electronic analogue computer should be seen as an alternative general-purpose computer, distinct from the digital domain, but equal in a broad range of applications, and even superior to the digital computer in a few.

R. J. Gomperts was one of the most outspoken critics of the bias towards digital computers and a keen advocate of the generality of the electronic analogue computer. In the early 1950s Gomperts was part of the design team at English Electric's Guided Weapons Division that developed LACE. LACE was a general-purpose electronic analogue computer designed to replace a number of one-off, single-purpose, analogue computer systems that had previously been built and used at English Electric's laboratories in Luton. In 1957 Gomperts reviewed the development of the LACE computer and stated:

> To summarize the arguments so far, we have tried to indicate that there is no reason why the predicate "general" should only be associated with digital machines. In fact, both types of machine, albeit using different techniques, simulate the mathematical model which is an abstraction of the physical problem.[12]

Gomperts' explanation for the apparent reluctance of the digital computer community, and others, to accept the new electronic analogue computer systems as general-purpose, was underpinned by the wider contemporary debate over the status of engineers.

> It is strange ... that analogue computers are considered in such a different light from the digital machines, seeing that they are intended to perform the same function ... The reason for this is not obvious but one might suggest that the digital machines were essentially developed for the use of mathematicians—extended desk machines—whereas the analogue machines have been mainly built by engineers for engineering purposes and [are] therefore *ipso facto* of a restricted nature.[13]

The connection that Gomperts was making explicit was that between the subordinate classification of analogue computers with respect to digital computing and the lower status of engineering *vis-à-vis* science. Electronic analogue computers were more often used in engineering applications than in scientific ones, and for analogue computer users the image of the analogue computer became entwined with images of the relative status of engineering and science. However, science versus engineering was not the only focus of contention. There was also intra-profession competition between the analogue and digital "men" within engineering.

The burgeoning digital computer industry had attracted large amounts of R&D funding from government and military agencies, as well as a great deal of public attention. Analogue computing, though expanding, had in fact declined in importance relative to digital computing. To an extent, so too did the status of those associated with it. It is perhaps not surprising then to find that analogue computer users played down images of the analogue computers as outdated, and as being "of a restricted nature", for it reflected upon their own professional status.[14]

The rapid pace of developments in digital computer technology and techniques was something that the analogue computer community could not ignore nor deny. It became increasingly difficult to maintain a directly adversarial position towards digital computing while retaining credibility. This brought about a shift in the debate. Instead of emphasising that analogue and digital computers were in direct competition, the analogue computer community negotiated a new position. This involved redefining the technologies as complementary rather than competitive. This was an astute change in the relations between analogue and digital computing, for it suggested that neither method was *a priori* "superior" to the other. It also attacked a hierarchical image of modern computing which placed the electronic digital computer at the top. Analogue computation could thus reassert itself as a distinct parallel development with its own non-subordinate hierarchical structure. One other important re-orientation in the analogue versus digital debate was to place greater emphasis on the circumstantial nature of any analogue/digital computer selection process. From this new position it followed that the "best" method—analogue or digital—could only be determined by a close inspection of the particular problem and a mix of other factors including speed and precision requirements, whether the computer was to be used in an "open" or "closed shop" environment, and the prior experience of those who were to use the equipment.[15]

Opening a discussion on developments in combined analogue and digital computation in Denver, Colorado, in 1956, John McLeod commented:

> Before we start the discussion I would like to make one thing clear: we do not intend to belabour the question of which method is better, analogue or digital. I hope that everyone present realizes that this is a question which can be answered only by intelligent consideration of what in heck you are trying to do; what are your inputs and what do you expect to do with the output; is much computation involved; and what are the speed and accuracy requirements?[16]

Thus, choosing to use an analogue computer rather than the digital alternative had become a decision that required "intelligent consideration", experience and even wisdom. Moreover, this choice was not to be interpreted as a declaration of the "restricted nature" of the application for which it would be used, nor of the person using it. Here we see the analogue computer community trying to distance itself from a directly adversarial "which method is better" view of analogue/digital computing.[17] Furthermore, the analogue computer community was using this contingent view of analogue/digital computing to negotiate an application domain for the analogue computer by emphasising that the computer selection process had to be application-specific. It thus appeared that the general-purpose electronic digital computer was not as "general-purpose" as it had been portrayed.

We can now look at two of the core issues in the debate: speed and accuracy/precision, and how these were employed in the negotiations for an application domain for electronic analogue computing.

DEFINING SPEED

In the analogue/digital debate, speed was problematised and differentiated. It came to be viewed as a parameter that was not easy to quantify or isolate from other computational factors, such as the nature and scale of the computation and the degree of precision required. One of the effects of the analogue/digital debate was that it highlighted the relative weakness of the digital computer in specific applications. Speed was contextualised. For the analogue computer community this meant that they were able to establish an application domain where electronic analogue computers had a speed advantage over digital systems. The analogue computer community used speed in the analogue/digital debate not only to help establish and maintain this application domain, but also to validate their claim that the digital computer was not *a priori* superior to the analogue.

From the very early stages of the analogue/digital debate, computing speed was a central issue. In the mid-to-late 1940s comparisons were based largely on the current performance, and expectations of the future performance, of electronic digital computers, *vis-à-vis* the mechanical differential analyzers. Two of the most influential digital computer pioneers, John von Neumann and Herman Goldstine, were among those who held and popularised the view that the main advantage of the electronic digital computer over previous and existing devices was extremely high computing speed. As William Aspray points out in his biography of John von Neumann, von Neumann and Goldstine believed that:

> The main mathematical significance of the electronic [digital] computer's speed was that it brought into range problems that were only marginally practical or entirely impractical to calculate using earlier devices: large runs of ballistic trajectories, astronomical orbit calculations, and parabolic and hyperbolic differential equations of fluid dynamics.[18]

In Britain, Douglas Hartree, a recent convert to digital computing, became one of its most ardent and influential supporters. In 1947, at his inaugural lecture as Plummer Professor of Mathematical Physics at Cambridge University, Hartree enthused about the advances in digital computing that were taking place in the USA, and in particular about the incredible speed of the ENIAC:

> The speed of the operation of the ENIAC is of the order of a thousand times faster than anything else at present available, and in any field of activity a change of scale of a factor of a thousand is difficult to appreciate.[19]
>
> ... Another example of its [the ENIAC's] speed is that it will do multiplication at the rate of about a million an hour, or say a hundred and fifty million a week, since it can work twenty-four hours a day and does not have to stop for food and sleep. Many calculations are not undertaken at present simply because of the sheer hard labour of carrying them out. But with the ENIAC and other electronic machines, a calculation involving ten million multiplications may only take about nine hours.[20]

Indeed, speed was of the essence, and though mechanical differential analyzers were fast compared with manual methods, compared with electronic digital methods they were not only slower, but they offered little scope for improvement. Thus, any such comparison left little doubt as to the superior performance of electronic digital methods. Yet in the USA by the late 1940s and early 1950s, the development of high-speed electronic analogue computers had tipped the balance back in favour of analogue computing. By 1950, general-purpose electronic analogue computers had been developed that could perform mathematical operations far faster than the existing mechanical differential analyzers. They were also faster than the contemporary electronic digital computers.

Yet for the purposes of the debate, defining the speed of a computer was not unproblematic, especially when comparing two fundamentally different methods of computing. The debate brought to the fore that though it was possible to compare how long it took to perform a single addition, multiplication or integration, this was not a very useful measure of relative performance. Real problems consisted of a great many combinations of these and other arithmetical and mathematical operations. One way to compare their relative speeds was to compare the time it took analogue and digital computers to solve complete problems. However, this too was problematic because different applications required different combinations of mathematical and arithmetic operations to be performed. Nevertheless, by the late 1940s and early 1950s in the USA and Britain, comparisons of electronic analogue and digital computers had in practice demonstrated that there was a category of mathematical problems for which electronic analogue computers did have a considerable speed advantage.[21]

It was generally agreed and recognised by advocates of both analogue and digital computing that the source of the comparatively high speed of operation of the electronic analogue computer was twofold: continuous data and parallel operation. In an electronic analogue computer, data vary and are operated on continuously. Mathematical operations such as integration, multiplication and function generation are carried out directly by separate computing elements, all of which are connected in parallel. Most of the contemporary digital computers had sequential, von Neumann architectures. In a digital computer, mathematical problems are solved by numerical methods which involve approximations to continuous functions. Integration and function generation require a large number of individual operations, each of which requires a finite amount of time. The speed of the digital computer was limited by this step-wise, serial mode of operation consisting of memory cycles, instruction execution, arithmetic operations and so forth. Taking the motor car as a metaphor for the arguments posited by the analogue computer community, it was not the speed of the engine that mattered but the speed of the car. Digital computers had fast engines, but there were a lot of numerical methods and a sequential transmission connected to it, that determined how fast the computer as a whole could run, or in other words its maximum road speed.

The fundamental differences between analogue and digital computing gave the analogue computer community considerable scope to negotiate over speed. For example, they were able to emphasise that one implication of a serial digital computer architecture was that the time digital computers took to solve a problem grew with the size of the problem. In the case of the electronic analogue computer, if the problem grew larger, the computer needed more computing elements. Thus the analogue computer grew physically, but since the computing elements were connected in parallel, the solution time remained the same. In 1955, W. W. Seifert, the Assistant Director of the Dynamic Analysis and Control Laboratory at MIT, argued the case for the electronic analogue computer as follows:

> Digital computers also can be employed for studying the mathematical model of a system. At the present time, the principal technical deficiency associated with their use in the study of complex systems lies in their speed. An analogue machine operates simultaneously on every term in a mathematical model of a system and thus requires no longer to solve a very complex problem than to solve a simple second-order equation. On the other hand, the time required to obtain a solution on a digital machine increases as the complexity of the problem studied increases. Even with the large, high-speed, electronic digital machines now available, the ratio of solution time on the digital machine to solution time on an analogue machine may run 50 to several thousand times for moderately complex system problems.[22]

Electronic analogue computer manufacturers were, not surprisingly, among those most eager to point out the nature of the speed advantage of the analogue computer. In 1955 John. D. Strong, who worked for Electronic Associates Inc., wrote:

> One important advantage of the analogue computer is its computing speed. As it performs all of its computing operations simultaneously (instead of sequentially as in the digital computer) the running time is independent of the size of the problem. On large problems the advantage is tremendous. One large analogue-computing laboratory, which has acquired a digital computer for checking purposes, reports a speed advantage on some of their problems of over 100 to 1.[23]

In the debate over the relative merits of analogue and digital computers, speed was enrolled by the analogue computer community as one of the enduring advantages of electronic analogue computing. However, speed could not be isolated from other computational parameters such as accuracy/precision. Accuracy/precision was not generally viewed as one of analogue computing's strengths. Nevertheless, the analogue computer community negotiated a non-pejorative view of the relative imprecision and inaccuracy of the electronic analogue computer, in the context of modern computing.

NEGOTIATING ACCURACY AND PRECISION

Initial comparisons of computational accuracy and precision seemed straightforward: electronic analogue computers offered only low-accuracy computation, and electronic digital computers offered high-precision computation. Yet by the early 1950s considerable confusion had arisen over the use of the terms "accuracy" and "precision"; this problematised attempts to compare analogue and digital computers. One of the sources of this confusion was that it had become common practice to refer to analogue computers as having an overall "accuracy" of, for example, 2%, 1% or 0.1%. Comparisons of the limited "accuracy" of the analogue computer with the theoretically infinite "precision" of digital computation left little doubt as to the superior performance of the digital method. Unfavourable comparison of analogue "accuracy" with the digital "precision" made this an important issue in the analogue versus digital debate.[24]

When this subject began to be addressed, the discussion usually began with formal definitions of "accuracy" and "precision", the intention being to demonstrate how they differed in the context of modern computing. The most commonly used definition of accuracy was "conformity to fact" or "truth". Based on this definition Albert S. Jackson, a former Assistant Professor of Electrical Engineering at Cornell University, further defined two components that contributed to the "conformity to fact" of a computer-derived solution:

1. Conformance of the model (equations) to the actual situation being studied.
2. Conformance of the operations of the computer to those called for by the model.[25]

Jackson's treatment of the accuracy/precision discussion typifies the contemporary debate from the perspective of the analogue computer community. The first of Jackson's two components of accuracy emphasised that the accuracy of a problem solution relied upon, and was delimited by, the accuracy with which the mathematical statement or electronic model of the problem conformed to fact. This definition of accuracy stressed the importance of understanding the nature of the problem, including the effects of underlying assumptions and approximations. Most importantly, it emphasised that there were other sources of inaccuracy and that they could be relatively large compared with those attributable to the computer. Moreover, quantifying these inaccuracies could be extremely difficult.

Jackson's second component referred to the loss of accuracy that was due specifically to the computer hardware and computing methods. Discussing this aspect of accuracy, the analogue computer community emphasised that both analogue and digital computers suffered from computational errors that affected their ability to conform to the operation of the model. Furthermore, error analysis was complex and involved the close examination of many aspects of a computer's operation. In a digital computer there were truncation errors to consider, as well as those which arose from sampling frequency, step rates and the fine-grain characteristics of integration algorithms. In the analogue computer some of the sources of computational error included component tolerances, non-linearities and bandwidth limitations in operational amplifier performance. Analysis was complicated by the fact that errors were not all cumulative, as some errors could cancel out others. However, one major weakness of the analogue computer that could not be denied was that as problems grew larger and more complicated, more computing elements were required, and the sources of errors increased as the number of components grew.

One other issue that was drawn into the accuracy debate was repeatability. Among the advocates of digital computing there was a widely held view that a computer's accuracy could be demonstrated by repeating a problem several times. If the same answer was obtained each time, then the operation of the computer was not contributing to inaccuracies. The fact that the digital computer could solve a problem a thousand times and give the same answer every time was thus taken to imply that it was inherently more stable and thus more accurate than the analogue computer. To an extent this was true, but the analogue computer community's reply was that this only demonstrated that it made the same errors a thousand times. On the other hand, the electrical noise in an analogue computer introduced a degree of random variation of parameters that sometimes led results to differ slightly. Yet though this may have seemed undesirable, there were those who made a virtue out of this inconsistency. They argued that these random variations tested the sensitivity of the mathematical/electronic model for operation in real-life conditions. This gave the problem originator greater confidence in the robustness of the final solution. This was something of a minority view, but it does help to indicate some of the scope for the negotiation and interpretation of accuracy.[26]

Fragmenting accuracy, and then discussing its components, enabled the analogue computer community to construct an interpretation of accuracy that demonstrated that digital computation could also exhibit low "accuracy". What was important from the perspective of the analogue computer community was that aspects of accuracy were viewed as independent of computing method, and that low accuracy was not necessarily unique to electronic analogue methods of computation.

The precision of computer-generated data was considerably easier to establish than its accuracy. On the question of the relative precision of analogue and digital computer solutions, there was no dispute that the digital computer had a considerable advantage. Digital computers could, if necessary, work with data that had thousands of significant figures. In theory, the solution provided by digital computers could have infinite precision. On the other hand, electronic analogue computers produced results that had a precision of only three or four significant figures.

It may have seemed that the community of analogue computer users and manufacturers had little room for negotiation. Nevertheless, it was able to construct a contingent view of precision that questioned the relevance of high-precision computation. It also established a more pragmatic attitude towards precision, based on the characteristics of individual applications. In doing so, the analogue computer community challenged, refuted and even ridiculed images of the infallibility of high-precision digital computation. Proponents of analogue computing argued that high-precision computation was often inappropriate, undesirable and could even be misleading. They promoted the view that the relatively low precision of electronic analogue computers was not *a priori* a serious disadvantage. The arguments marshalled by the analogue computer community in defence of the relatively low-precision of analogue computer centred on cost and the relatively low precision of input data.

General-purpose electronic analogue computers in the 1950s were generally less expensive than digital systems, and were usually bought outright rather than being hired. The analogue computer community argued that for applications where the relatively low precision of the analogue computer was acceptable, the additional cost of the digital computer was not justifiable. However, as smaller and less expensive digital computer systems became available from the mid-1950s, it became increasingly difficult to sustain this argument, especially since the digital computer could be used for a wider variety of applications.[27]

The main thrust of the argument of the analogue computer community was that high-precision computing should not be valued for its own sake. Proponents of the analogue computer emphasised that in the real world much of the data that engineers and others worked with was low-precision data.[28] Examples cited included test data from wind tunnels and field trials, which was usually measured with much less precision than an analogue computer was capable of working to. Commenting on the relationship between the problem, its data, and computational precision, Albert S. Jackson, noted:

> The analog computer is not a precise calculating machine in the sense that the modern digital computer is. The variables are continuous quantities rather than precise numbers. The answers are usually measured to only three or four decimal digits; therefore, the analogue computer is restricted to problems either where the answer is not required to a higher degree of precision than this or—the more common case—where the input data are actually not known to any better precision. Obviously, if the parameters of a problem are not known more precisely than to three significant figures, an answer carried out to twelve significant figures and used as such, is little short of ridiculous.[29]

In this view, computational precision was contingent and application-specific. Taking cost and precision together, proponents of analogue computing, including the manufacturers, were saying— "Why pay for extra precision if you don't need it?" Moreover, users should realise that the degree of precision that they require from their computers should be determined by consideration of the problems in hand, and the nature and precision of the input and output data. Almost reproachfully, users are reminded that the inappropriate use of high-precision data is "little short of ridiculous".

Notwithstanding the rhetorical components of such discourse, there was also a belief among analogue computer users (usually engineers) that too much faith was being placed in digital computer solutions, which were expressed as an impressively long line of digits. Jackson, was among those who encouraged greater scepticism towards the merits of high-precision digital computation:

> It should be emphasized that, although a numerical answer may result very precisely and accurately from operations performed on precise input numbers, it may yet give an erroneous picture of the physical situation being studied. A numerical result expressed in many digits at the end of a computation is regarded by many people as an indisputable truth.[30]

Thus it was being argued that the "indisputable truths" of high-precision numerical results could be misleading. They were at the very least a long way from the earthy imprecision of the oscilloscopes and electronic models of the analogue computer. Indeed, the fallibility of the electronic analogue computer was in this view an inherent asset, because it required users, often design engineers, to maintain a healthy scepticisim at all time.

ACCESS AND INTERACTION: COMPARING ANALOGUE AND DIGITAL COMPUTER ENVIRONMENTS

The debate was not limited to technical aspects of the computers themselves, but also addressed organisational aspects of the environments in which the computers were used. While analogue computing was organised on an "open shop" basis, digital computing took place in a "closed shop" environment. In the "closed shop" system the problem originator was seldom permitted access to the computer. Instead, the problem was submitted to a computer operator and entered a queue containing other "jobs"; the problem was "run" sometime later when the operator or computer allocated the problem computer time. The results, which generally appeared in numerical form, were then handed back to the person who submitted the problem for analysis. If changes were required, this time-consuming procedure had to be repeated.

The "closed shop" environment is mentioned by M. V. Wilkes as an important reason why digital computers did not have "an early and significant impact" on engineering design. In 1967 Wilkes presented a paper at a symposium on "The Impact of Digital Computers on Engineering", organised jointly by the Royal Aeronautical Society and the Institution of Mechanical Engineers. He argued that, at first, digital computers were too small to be useful for engineers. He continued:

> Unfortunately, the big computers, when they came, were so expensive that it was not considered economic to allow users to handle them personally, and batch-processing techniques were introduced. The result was to create a barrier between the computer and the design engineer, and to make it impossible for him to get results of any kind without a delay amounting to a few hours at the very best, and often to much more. Emphasis, in fact, was put on the efficiency with which the central processor of the computer was used and no regard at all was paid to the efficiency with which the user—in this case programmers and design engineers—worked.[31]

Analogue computers were in contrast generally operated on an "open shop" basis, with the setting-up and operation of the computers undertaken by the same person who originated the problem; results appeared immediately and changes could be made easily and without delay. The proponents of analogue computing argued that the ease of access of the "open shop" environment stimulated better engineering design by allowing continuous interaction. In 1955, W. W. Seifert wrote:

> If the engineer who is analyzing the system follows the study through its various stages and takes actual responsibility for operating the analogue computer, then he can apply all his knowledge of the operation of the system in making on-the-spot decisions concerning the parameter values that should be studied. From his first-hand observations of difficulties with the analogue setup, he may be able to predict weaknesses that would arise in the analogous physical system.[32]

Thus, in the early 1950s analogue computing offered access and interaction, while digital computing was associated with a type of organisation that maintained a distance between the engineer and the problem.

RENEGOTIATING FOR HYBRID COMPUTERS

Hybrid computers were to an extent an explicit recognition of the relative strengths and weaknesses of electronic analogue and digital computers. In the late 1950s the growth of interest in hybrid computers diffused much of the adversarial view of analogue versus digital computing, and helped to re-enforce an image of them as complementary rather than competitive technologies. This was particularly true for the category of hybrid computer known as "true hybrids", which were the combination of a general-purpose analogue computer with a general-purpose scientific digital computer, in a synergistic alliance. It was argued that hybrid computers combined the best of both worlds: the high speed of the analogue computer with the high precision and programmability of the digital computer. Nevertheless, though hybridisation brought many benefits, not everyone welcomed the general trend. The emergence of hybrid computing thus brought about a reorientation in the debate. From the late 1950s the analogue/digital debate had two new foci: the comparison of hybrid versus all-analogue computing and the comparison of hybrid versus all-digital computing.

All-analogue versus hybrid computing

The principal concern of the all-analogue versus hybrid computer debate was the question of the relative merits of interactive analogue computing versus remote, batch-processing digital computing methods. As we have seen, this was also an aspect of the earlier analogue/digital computer debate. In this part of the debate, the emphasis was on the increasing use of the digital computer to control the analogue computer's operation. Much of the discussion was internal to the analogue computing community. The main concern of the analogue computer community was how this digital computer interface would affect the interactive relationship between the computer user, the problem and the analogue computer model.

One side of the argument was that the digital interface isolated the problem originator from the electronic/mathematical model set up on the analogue computer. The imposition of digital interfaces was seen as something of a betrayal of a basic tenet of electronic analogue computation: the implicit and explicit benefits of hands-on, interactive computing, including the routine process of manually setting up and checking out the computer. Yet the process of setting up and checking the analogue computer could be extremely tedious and time-consuming, especially for large problems. Thus automation helped to make analogue computers more accessible to newcomers, and enhanced their flexibility, efficiency and reliability. Hybridisation also offered benefits in terms of programme storage, data gathering and manipulation that could augment all-analogue computing.

Notwithstanding these benefits, digital control of the analogue computer had several negative implications for the analogue "men". It meant becoming conversant with digital programming techniques and thus, to an extent, crossing over to the realm of the digital "men". Automation also meant taking some of the "art" out of the "analogue art". For a generation of electronic analogue computer experts, who had spent many years developing the very specific skills that analogue computing required, hybridisation meant technical change, but not necessarily progress. Stalwarts of all-analogue computing were left lamenting the loss of, what in their view was, a quintessential and irreplaceable aspect of analogue computation. Yet most accepted that no amount of renegotiation, based solely on the tacit benefits of all-analogue computing, could maintain the boundaries of an application domain which the proponents of hybrid, and all-digital, computing were trying to establish as their own.

Hybrid versus all-digital computing

From the mid-1960s, with the introduction of the first commercial true hybrid computer systems, the analogue/digital debate was effectively superseded by the hybrid versus all-digital computer debate. The rapid and systematic developments in all aspects of digital computing, including components, architectures and software, led to improvements in speed, accessibility and lower hardware costs. One effect of the phenomenal growth and diversification of the digital computer industry was that the area of overlap between the application domains of all-digital and hybrid computers increased, at the expense of the hybrid computer. By the early 1970s, it was becoming increasingly difficult for hybrid computing to compete with the giant digital computer industry and to maintain a distinct application domain.

The hybrid computer industry felt that much of its lack of success arose from an irrational and aggressive campaign by the proponents of all-digital computing to have the hybrid computer declared obsolete. Writing in 1973, R. Bubloz of the leading American hybrid computer firm Electronic Associates Inc., commented:

> The motor car has enjoyed a growth similar to that of the digital computer. Most families in the western world own a car; most engineers and scientists have access to a digital computer. But no sensible person would argue that because they owned a car, other forms of transport would soon become obsolete. Surely, when planning a journey we weigh up factors such as time available, economics and distance, to determine the best way of travelling. In the same way it is prudent to consider similar factors when embarking on a computational problem.[33]

Proponents of hybrid computing argued that the success of the all-digital computer was largely due to a lack of appreciation of the relative merits of all-digital and hybrid computing. Rallying against the competition from digital computing, hybrid computer manufacturers argued that engineers and scientists were unable to evaluate the available options and, that they were unable to do so because of a lack of training, laying much of the blame for this situation at the door of the educational establishments.[34] To quote again from R. Bubloz:

> relatively few people ... are trained to appreciate the relative merits of these machines compared with conventional digital processors.
> ... The attitudes prevailing today have been brought about by the somewhat one-sided bias of the many educationalists who have become infatuated with the digital computer.[35]

For the proponents and manufacturers of hybrid computers there was little space left, in which to renegotiate a place and a role for hybrid computers. Without training and an understanding of the operation of hybrid, as well as digital, computing techniques, and without exposure to the latest developments, potential users would neither appreciate the advantages of hybrids nor choose to use them. Yet if there was to be a future for hybrid computing, it relied upon maintaining the maxim that computer selection was contingent, contextual and should be based upon informed choice.

From the point of view of the proponents of hybrid computing, the main advantages that potential users should have been made aware of were the higher computational speeds and cost savings that hybrid computing could bring. In support of these claims, advocates of hybrid computing undertook studies to demonstrate the efficacy of hybrid computers in real-time simulation studies and hardware-in-the-loop applications. The argument for cost savings was made on the basis of applications characterised by high complexity, and where the calculation had to be performed a great many times. The speed advantage of the analogue part of the hybrid computer led to considerable savings in computer time, and thus cost.[36] Yet even though such benefits were demonstrable, hybrid computers were superseded by all-digital ones, the hybrid versus all-digital debate was marginalised, and closure followed.

CONCLUSION

Throughout the 1950s and 1960s the manufacturers of analogue/hybrid and digital computers competed to establish and maintain their share of the market in computers for engineering and scientific applications. The analogue computer industry strove to secure a place and a role for analogue/hybrid computers in the sphere of modern automated computation. Yet though many of those actively promoting the use of analogue/hybrid computation came from the analogue/hybrid computer industry, commercial interests were not the only force driving the debate. From the late 1940s, firms, colleges, universities and research organisations invested in computer hardware, and users invested significant amounts of time and effort to obtain the intellectual software, or "know-how", specific to electronic analogue computation. By the mid-1960s many people had spent upwards of fifteen years working in the analogue computer domain, developing and/or using electronic analogue computer systems. Thus proponents of analogue computing defended it, because it was their chosen specialisation. Capital, as well as personal, investment generated both technological momentum and a degree of inertia against change.

Though this goes a long way to explaining why analogue "men" were eager to sustain their art and were so reluctant to cross over to the digital domain, two explanatory factors need restating. The first is that in their own view it was neither clear nor inevitable, that the digital computer would supersede the electronic analogue computer. Indeed, newcomers chose to specialise in analogue computing for many years after the advent of digital computer systems, and not only because the departments they joined did not have digital computing equipment. The second is that electronic analogue computers not only embodied, but also enhanced traditional engineering practices. The commitment of engineers to postwar electronic analogue computation reflected their commitment to a largely pre-World War II engineering culture and pedagogy that stressed empirical methods and visualisation over theory and analytical techniques. Negotiations to sustain an application domain for analogue computation can also be interpreted as negotiations for a domain where "doing" and "know-how" continued to be valued higher than "knowing".

Yet this is not to say that electronic analogue computers were not used for, or applicable to, scientific investigation, or that digital computers were not widely used and extremely useful in engineering design. However, the primary context and *raison d'être* for electronic analogue computers was the environment of the engineering design laboratory. When engineers chose to become either analogue or digital "men", it was not necessarily an irreversible choice. The widespread adoption of hybrid computers in the 1960s made the boundary less clear, but many from both communities retained their commitment and remained sceptical of the claims of superior performance made by the advocates of the alternative competitive technology.

NOTES

1. *Simulation*, Aug 1965, p. 101
2. An exception to this is the book by P. N. Edwards, *The Closed World: Computers and the Politics of Discourse in Cold War America*, MIT Press, Massachusetts, 1996, see chapter 2, pp. 66–73
3. W. G. Vincenti, "The Air-Propeller Test of W. F. Durand and E. P. Lesley: A Case Study in Technological Methodology," *Technology and Culture*, Vol. 20, no. 4, 1979, pp. 712–751, p. 742.
4. Ibid., p. 742.
5. E. T. Layton, Jr., "Scientific Technology, 1845–1900: The Hydraulic Turbine and the Origins of American Industrial Research", *Technology and Culture*, Vol. 20, no. 1, 1979, pp. 64–89, p. 88. Here Layton states that "The world of engineering is not an ideal mathematical world at all, but something more complex and qualitatively distinct. Engineers use theory, but a tedious process of development involving testing and scaling up in essential. To date, the complexities of practice continue to transcend theory. Engineering science is teleological; like all technological activities, it serves social ends exterior to itself, which are manifest in design." p. 89.

6. D. B. Breedon, "Analog versus digital techniques for engineering problems", *Trans., IRE*, IE-4, 1957, pp. 86–89.

7. Anon., "Digital vs. Analog Computers", *Simulation Council News Letter*, in *Instruments & Control Systems*, Vol. 32, May 1955, p. 803–806.

8. R. B. Quarmby, "Electronic Analogue Computing: Survey of Modern Techniques", *Wireless World*, March 1954, pp. 113–118, p. 113.

9. L. Cahn, quoted in J. McLeod, "Ten Years of Computer Simulation", *Trans., IRE, Electronic Computers*, EC-11, Feb 1962, pp. 2–6, p. 3.

10. Arguably the most notable early work on the concept of a "universal" machine was that undertaken by the British mathematician Alan Turing in the mid-1930s. In a paper published in 1936, Turing indicated that a single machine could be capable of performing any logical operation that could be mechanized, the implication being that any number of special-purpose computing machines could be replaced by a general-purpose one. A. M. Turing, "On Computable Numbers, with an Application to the Entscheidungsproblem", *Proc., London Mathematical Society*, Vol. 42, 1936, pp. 230–265; A. Hodges, *Alan Turing: The Enigma of Intelligence*, Burnett Books and Hutchinson, London, 1983, pp. 102–110.

11. W. Allison, "A Broad Look at Analog Computers", *Control Engineering*, February 1955, pp. 53–57, p. 54.

12. R. J. Gomperts and D. W. Righton, "LACE: The Luton Analogue Computing Engine, Pt 1", *Electronic Engineering*, Vol. 29, July 1957, pp. 306–312. p. 307.

13. Ibid., p. 307; see also R. J. Gomperts, Computing applications where analogue methods appear to be superior to digital", *Journal Brit., IRE*, Aug 1957, pp. 421–428.

14. R. A. Buchanan, *The Engineers: A History of the Engineering Profession in Britain, 1750–1914*, Jessica Kingsley, London, 1989; R. Friedel, (ed.), Special Issue on "Engineering in the Twentieth Century", *Technology and Culture*, Vol. 27, no. 4, 1986; E. T. Layton, Jr., *The Revolt of the Engineers: Social Responsibility and the American Engineering Profession*, Johns Hopkins University Press, Baltimore, 1986; P. Meiksins, "The 'Revolt of the Engineers' Reconsidered", *Technology and Culture*, Vol. 29, No. 2, 1988, pp. 219–246; D. Noble, *America by Design: Science, Technology, and the Rise of Corporate Capitalism*, Oxford University Press, Oxford, 1977.

15. *Simulation Council News Letter*, in *Instruments & Automation*, Vol. 28, July 1955, p. 1131–1135.

16. J. McLeod, *Simulation Council News Letter*, in *Instruments & Automation*, Vol. 29, Oct 1956, pp. 2015–2023, p. 2015.

17. H. Meissinger, "Trends in Simulation", *Simulation Council News Letter*, in *Instruments & Control Systems*, Vol. 32, Sept 1959, pp. 1389–1393.

18. W. Aspray, *John von Neumann and the Origins of Modern Computing*, MIT Press, Cambridge, Mass., 1990, pp. 60–61.

19. D. R. Hartree, *Calculating Machines: Recent and Prospective Developments, and their Impact on Mathematical Physics*, Cambridge University Press, Cambridge, 1947, p. 24.

20. Ibid., pp. 26–27.

21. This category consisted largely of problems associated with the solution of sets of differential equations. Electronic analogue computers could compute the solution to such problems in real time, or a fraction of real time. This combination made them well suited to simulation studies of dynamic systems.

22. W. W. Seifert, "The Role of Computing Machines in the Analysis of Complex Systems", Paper presented at the International Analogy Computation Meeting, Brussels, Sept 26–Oct 1, 1955, pp. 1–19. *Typescript*, MIT Archives, MC6, p. 10.

23. J. D. Strong, "A Practical Approach to Analog Computers", *Instruments & Automation*, Vol. 28, April 1955, pp. 602–610, p. 610.

24. L. N. Ridenour, "The Role of the Computer", *Scientific American*, Vol. 187, 1952, pp. 116–130.

25. A. S. Jackson, *Analog Computation*, McGraw-Hill, New York, 1960, p. 6.

26. Personal Interview: John McLeod, La Jolla, Ca., 11–12 October 1990.

27. For example, the Bendix G-15 was the first small scientific computer it became available in 1956, cost approximately $45,000, and more than 300 were sold. The same year a small scientific computer was also introduced by Librascope—the LGP-30; more than five hundred of these were built. See, K. Flamm, *Creating the Computer: Government Industry and High Technology*, Brookings Institution, Washington DC, 1988, pp. 66–67, and p. 75.

28. C. B. Crumb Jr., "Engineering Uses of Analog Computing Machines", *Mechanical Engineering*, Vol. 74, Aug 1952, pp. 635–639, p. 639.

29. A. S. Jackson, *Analog Computation*, p. 6.

30. A. S. Jackson, *Analog Computation*, p. 7.

31. Historically, Wilkes is chiefly known for his work on the development of the EDSAC digital computer at Cambridge University in Britain in the late 1940s and 1950s. M. V. Wilkes, "The Coming of Multiple-Access Computers", *Journal of the Royal Aeronautical Society*, Vol. 71, no. 676, April 1967, pp. 235–236, p. 235.

32. W. W. Seifert, "The Role of Computing Machines in the Analysis of Complex Systems", pp. 10–11.

33. R. Bubloz, "Appreciating the analogue and hybrid techniques", *Computer Weekly*, 30 Aug 1973, pp. 6 & 12, p. 6.

34. R. W. Olmstead, *Simulation Council News Letter*, in *Instruments & Control Systems*, Vol. 36, Oct 1963, p. 117; O. Serlin, *Simulation Council News Letter*, in *Instruments & Control Systems*, Vol. 36, Feb 1963, p. 142.

35. R. Bubloz, "Appreciating the analogue and hybrid techniques", p. 6.

36. V. J. Sorondo, R. B. Wavell and F. E. Nixon, "Cost and Speed Comparisons of Hybrid VS. Digital Computers", paper presented at, the *Special Symposium on Advanced Hybrid Computing*, sponsored by US Army Materiel Command, *Summer Computer Simulation Conference*, San Francisco, Ca., 24 July 1975; L. Wolin, "The Wolin–Saucier–Peak (WSP) Scientific Mix: A Quantitative Method for Comparing Hybrid and Digital Computer Performance", paper presented at, the *Special Symposium on Advanced Hybrid Computing*, sponsored by US Army Materiel Command, *Summer Computer Simulation Conference*, San Francisco, Ca., 24 July 1975.

CHAPTER 9

Conclusion

The origins of electronic analogue computing lie in a tradition of computing by analogy that can be traced back to ancient times. However, though entirely consistent with this tradition, the origins of the post-World War II general-purpose electronic analogue computer do not lie in the major prewar analogue computing devices that are generally portrayed as its predecessors, namely the general-purpose mechanical differential analyzers and electrical network analyzers. As we have seen, the electronic analogue computer was not the product of a systematic, rational programme to progressively apply electronics to analogue computing devices already widely used by scientists and engineers as aids to computation and design. The electronic analogue computer did depend upon prior technology and techniques, but the origins of the key technology, the DC amplifier, lay not in analogue computing, but in the development of amplifiers for telecommunications and radio and automatic control applications. The process of enrolment and redefinition of thermionic-valve-based amplifiers as analogue computing components, which began in the late 1930s and 1940s, was largely independent of the major prewar centres of analogue computer development in academia, and was undertaken by different individuals for different purposes. Moreover, the commercial electronic analogue computer systems which emerged in the USA in the late 1940s and in Britain in the early 1950s were the product of industrial, rather than academic, research and development. Thus, the electronic analogue computer emerged as a concurrent alternative to the other major analogue as well as digital computers, rather than as a replacement for them.

In this book I have argued against approaches that ignore or marginalise technological alternatives. The history of the emergence of the electronic analogue alternative demonstrates and emphasises the existence of technological parallelism and discontinuity, and thus constitutes an explicit critique of uni-linear, "trajectory"-based models of technical change. It is also a critique of historical perspectives which foster such ahistorical models by taking a dominant technology from the present and seeking to explain its development in terms of an unbroken line of precursors that lead inevitably to its most modern and highly developed form. Until recently, this Darwinian view of technological change has been the dominant perspective in the history of computing literature. In this view all previous analogue and digital computing devices are portrayed as mere precursors to the modern electronic digital computer. In this well-ordered progression, the superiority of the postwar electronic digital computer is, necessarily, unequivocal. Yet as we have seen, there were two alternative postwar computing technologies: the electronic analogue computer and the electronic digital computer. Nevertheless, the electronic analogue computer has been almost entirely neglected historically. This, I suggest, is partly because it constitutes a discrepancy to the evolutionary view of technical change in that it could be portrayed neither as a prelude to the electronic digital computer nor as a step in that direction. Furthermore, it could not be portrayed as a step in another direction that rapidly led to failure.

In analysing this history we see that no single pioneering story provides a channel through which the economic, institutional or technical forces, that shaped electronic analogue computing or the industry can be viewed. What we find instead is a developmental process involving technical and economic imperatives, military agencies, civilian and government bodies, commercial companies, universities, private firms and research institutes. Electronic analogue computer development was interwoven with developments in electronic devices, aerospace and digital computing technologies. Electronic analogue computers were developed largely by engineers for engineers and embodied traditional forms of engineering practice, including graphical, trial-and-error, parameter-variation and model-building methods.

In Britain and the USA, electronic analogue computers were developed in response to both the increasing complexity of technical systems and the impracticality and inadequacy of existing analytical and empirical design methods. In particular, the computational demands associated with the development of guided missiles and aircraft stimulated interest in, and provided the principal *raison d'être* for, the development of electronic analogue computers. As we have seen, economic factors associated with the development of weapon systems encouraged military and government bodies to fund research and research and development work on electronic analogue computing. This work was pivotal and established much of the technological and institutional basis for the subsequent commercialisation of general-purpose electronic analogue computers. For more than three decades military patronage fostered and sustained developments in electronic analogue and hybrid computing.

Yet if as Larry Owens suggests, technologies can be read as texts, then reading the electronic analogue computer does more than to point up the pervasive influence of the military-industrial complex on the course of technological change.[1] The electronic analogue computer can also be read for insights into the values, practices and dynamics of engineering, as well as how it differs from science. As we have seen, neither economic nor technical factors alone can fully explain the success and/or appeal of electronic analogue computers, nor can they explain the commitment of their proponents to them. Electronic analogue computers were shaped by and resonated with, a largely pre-World War II engineering culture that valued "know-how" more than "knowing", and continued to view engineering as more art than science. Yet as Eugene Ferguson has pointed out, since the end of World War II,

> The art of engineering has been pushed aside in favour of the analytical "engineering sciences", which are higher in status and easier to teach.[2]

The debate between the analogue and digital computer communities can therefore be read as part of a wider debate about professional status. We see that this overarching debate concerned not only the relative status of engineering versus science, but also that of the "art of engineering" versus "engineering science". I suggest that those who defended the electronic analogue computer against charges of outdatedness and anachronism *vis-à-vis* the digital computer were

simultaneously defending the status of traditional engineering practices and values *vis-à-vis* science and modern "engineering science".
Indeed, not to have done so would have been tantamount to admit-
ting that there was no longer a place in engineering for knowledge
that could not be expressed mathematically, or a role for "intuition"
and a "feel" for the problem in engineering design.

Yet the use of electronic analogue computers led not only to the
growth of tacit forms of technological knowledge but also to the
revision of engineering theory. They enabled engineers to build active
models that embodied and operationalised the mathematical symbol-
ism of engineering theory. With these models, empirical methods
could be used to study the behaviour of a technical system beyond
the limits predicted by extant theory. Electronic analogue computers
helped designers to develop increasingly complex technical systems
by bridging the gap between the limits of earlier empirical methods,
current analytical techniques and the real-world systems that they
were constructing. In doing so, the electronic analogue computer
became, somewhat ironically, a victim of its own success. More complicated technical systems required larger and more detailed models;
because of the parallel and direct nature of the electronic analogue
computer, as models grew, so too did the number of computing elements that had to be interconnected. Setting up a problem of a
hundred computing elements required several hundred interconnections to be made and checked. Electronic analogue computers not
only became increasingly difficult to use, but it also became increasingly difficult for the user to get a "feel" for the problems being
studied. Systems involving the use of digital computers were developed to automate the process of setting up, checking and component
interconnection (or patching). But as we have seen, while overcoming
some of the problems created by increasing complexity, the digital
interface tended to alienate the user from the analogue computer
setup. Electronic analogue computers became less "engineering-
friendly", and their ability to operate as a direct and unique aid to
the design process diminished.

The development of the electronic digital computer ran parallel to
that of the electronic analogue computer, but was undertaken on a
much greater scale. The market for digital computers for use in business—in finance, data management and logistics applications—grew

rapidly from the early 1950s. As Edwards has pointed out, among the many forces driving forward the development of the digital computer, was the promise of creating "Artificial Intelligence". This created a nexus of interest groups from universities, business and the military that spurred on development work and provided funding.[3] An enormous amount of research and development effort was directed towards the improvement of all aspects of digital computing hardware and software. The combined endeavour led to the development of faster central processing units, larger-capacity memories, and scientific programming languages. The development of time-sharing and the introduction of direct on-line digital computing led to organisational changes that brought about improved access for digital computer users. In combination, these changes meant that digital computers became significantly more user-friendly than their predecessors had been.

Thus, while analogue computers lost some of their appeal to engineers, digital computers were becoming more attractive. The differences between the two technologies were reduced: whereas analogue computers became increasingly dependent on digital technology to manage complex problems, digital computers became directly accessible and could be operated much like the analogue computers, i.e. on an "open shop" basis. Technological and organisational convergence shifted the debate about the relative merits of analogue and digital computers away from qualitative issues, towards quantitative ones of technical and economic performance. In this respect, analogue and hybrid computers could claim greater speed and therefore an economic advantage in a particular category of problems. Digital computers, however, were far more versatile and could be employed in a much wider range of applications, including those beyond the immediate concerns of engineers and scientists. They had therefore become much more widely adopted than either electronic analogue or hybrid computers. Digital computing thus benefited from what Brian Arthur has referred to as "increasing returns to adoption", whereby one of a number of competing technologies becomes more attractive, the more it is adopted.[4] Where it was indeterminate whether an analogue/hybrid computer or digital computer was better for a particular application, the more widespread adoption of the digital computer mitigated in its favour, but so too did its extra versatility. In his study of the replacement of spark by continuous wave radio in the USA between 1900 and 1932, Hugh G. J. Aitken posits that one technology is able to displace another when:

> It does more than merely perform old functions better; it makes it possible to perform functions that the technology it replaces could not perform at all. It not only solves a problem; it "over-solves" it. It literally creates its own future.[5]

Digital computers did not perform all functions better than electronic analogue/hybrid computers, but by the late 1970s they were able to perform most functions equally well, and their versatility meant that they could also perform many other functions. To this extent they did indeed "over-solve" the problem. However, as many studies have already shown, a technology neither creates itself nor "its own future". The digital computer did not autonomously develop into a form that was capable of supplanting the electronic analogue/hybrid computer. Its development was shaped by many of the factors that also influenced the development of the electronic analogue computer, and this development required a substantial investment in terms of both human endeavour and capital. Moreover, as this study has shown, the rise to dominance of digital computing technology was by no means universally regarded as a foregone conclusion. Indeed, this study reveals that there was a much more robust belief in the future of the electronic analogue alternative than has so far been acknowledged in the historical literature on computing.

To conclude, many factors, both internal and external, played a role in shaping the development, and subsequent decline, of postwar electronic analogue computing in Britain and the USA. The history of the electronic analogue computer and the debate about the relative merits of postwar electronic analogue and digital computing have demonstrated the equivocal and contingent nature of postwar computer development. We see that neither the success nor the failure of alternative technologies is pre-ordained.

NOTES

1. L. Owens, "Vannevar Bush and the Differential Analyzer: The Text and Context of an Early Computer", *Technology and Culture*, Vol. 27, no. 1, 1986, pp. 63–95.
2. E. S. Ferguson, *Engineering and the Mind's Eye*, MIT Press, Cambridge, Mass., 1992, see Preface.

3.   P. N. Edwards, *The Closed World: Computers and the Politics of Discourse in Cold War America*, MIT Press, Massachusetts, 1996 see chapter 8

4.   W. B. Arthur, "Competing technologies: an overview", in G. Dosi *et al.*, (eds.), *Technical Change and Economic Theory*, Pinter, London, 1988, pp. 590–607, p. 590.

5.   H. G. J. Aitken, *The Continuous Wave: Technology and American Radio 1900–1932*, Princeton University Press, Princeton NJ, 1985, p. 11.

Bibliography

Abdank-Abakanowicz, *Die Intergraphen*, B. G. Teubner, Leipzig, 1889.

Adams, A. R., *Good Company: The Story of the Guided Weapons Division of British Aircraft Corporation*, British Aircraft Corporation, Stevenage, 1976.

Adams, B. D., *Ballistic Missile Defence*, American Elsevier, New York, 1971.

Adams, W. G., *Proceedings of the Royal Society*, Vol. 24, 1876, p. 1.

AEI 955 Analogue Computer, Trade Publication, Associated Electrical Industries, Trafford Park, Manchester, May 1960.

Aitken, H. G. J., *The Continuous Wave: Technology and American Radio 1900–1932*, Princeton University Press, Princeton NJ, 1985.

Allison, D. K., "US Navy Research and Development since World War II", in M. R. Smith, *Military Enterprise and Technological Change: Perspectives on the American Experience*, MIT Press, Cambridge, Mass., 1985.

Allison, W., "A Broad Look at Analog Computers", *Control Engineering*, Feb 1955, pp. 53–57.

Anon., "A General Purpose Electronic Analogue Computer", *The Engineer*, 25 Sept 1953, pp. 395–397.

Anon., "A Nucleonics Survey: Analog Computer Use in the Nuclear Field", *Nucleonics*, Vol. 15, no. 5, 1957, p. 88.

Anon., "An Automation Progress Survey: Principal British Analogue Computers", *Automation Progress*, Nov 1958, pp. 403–405.

Anon., "Analogue Computer: Capable of Expansion," *Engineering*, Vol. 191, 7 Sept 1962, p. 307.

Anon., "Analogue Computer: Educational," *Engineering*, 1 Oct 1961, p. 439.

Anon., "Analogue Computer", *Engineering*, Vol. 188, Oct 1959. p. 476.

Anon., "Analogue Computers", *The Computer Bulletin*, Vol. 3, Dec 1959, p. 76.

Anon., "Applications of the RADIC System", *Typescript report*, Redifon Ltd, Crawley, c. 1960.

Anon., "Atomic Review: Pattern of Research", *Engineering*, Vol. 182, Aug 1956, pp. 283–285.

Anon., "Biographies of Significance: Stanley Rogers", *Instruments & Automation*, Vol 28, Nov 1955, p. 1847.

Anon., "British Computers Show their Paces", *Automation Progress*, Nov 1958, pp. 406–411.

Anon., "Characteristics of General Purpose Analog Computers", *Computers and Automation*, Vol. 13, June 1964, p. 78.

Anon., "Characteristics of General Purpose analog and Hybrid Computers," *Computers and Automation*, Vol. 18, Computers," 1969, pp. 151–183

Anon., "Computers and Computing at the City University", *Typescript report*, City University, London, September 1981.

Anon., "Concorde Flight Simulation at Filton", British Aircraft Corporation, Filton Division, Engineering Computing Services, *Typescript report*, October 1972.

Anon., "Development of the electrical director," *Bell Laboratories Record*, Vol. 22, no. 5, 1944, pp. 225–230.

Anon., "Differential Analyzer at Manchester University", *Engineering*, Vol. 140, 26 July 1935, pp. 88–90.

Anon., "Differential Analyzer: The N.P.L.'s New Analogue Computer for Solving Differential Equations", *Engineering*, Vol. 178, 19 Nov 1954, pp. 659–660.

Anon., "Differential Analyzer", *The Engineer*, Vol. 160, July 26, 1935, pp. 82–84.

Anon., "Digital vs. Analog Computers", *Simulation Council News Letter*, in *Instruments & Control Systems*, Vol. 32, May 1955, pp. 803–806.

Anon., "EAI Computation Centre in Los Angeles Serves West", *Western Electronic News*, December 1956, p. 1–4.

Anon., "Electronic Computer Exhibition", *Process Control and Automation*, Vol. 5, no. 11, Nov 1958, pp. 487–496.

Anon., "Firth Cleveland Back Solartron Group," *Engineering*, 15 Jan 1960, p. 89.

Anon., "Hybrids and Analogues" *Computer Weekly*, 11 Jan 1968, p. 1.

Anon., "Improved Analogue Computer Trainer," *Engineering*, Vol. 196, 27 Dec 1963, p. 812.

Anon., "Miniature Analogue Computer", *Engineering*, Vol. 179, May 1955, p. 659.

Anon., "New RCA Electronic Computer Aids US Air Defence;...". RCA, New York NY, 2 Nov 1950, pp. 1–5.

Anon., "Operational Flight Trainer Uses 200 Tubes," *Electronics*, Feb 1945, pp. 214–216.

Anon., "Oxbridge Computes with the Times," *Engineering*, 8 June 1962, p. 761.

Anon., "PACE Computers", *The Computer Bulletin*, March 1963, p. 150.

Anon., "Principal British Analogue Computers", *Automation Progress*, Nov 1958, pp. 404–405.

Anon., "RCA Computer Aids Missile Design", *Product Engineering*, Jan 1951, p. 18.

Anon., "Reactor Simulators," 23 June, 1961, *Engineering*, p. 864.

Anon., "Reactor Simulators", *Engineering*, 23 June 1961, p. 864.

Anon., "Research Leadership at United Aircraft and Beckman", *Instruments & Control Systems*, Vol. 35, 1962, p. 170.

Anon., "Simulating Helicopter Flight," *Engineering*, 18 Oct 1957, pp. 500–501.

Anon., "Solartron SC 30 Analogue Computer," *Engineering*, Vol. 191, 23 June 1961, pp. 874–876.

Anon., " Survey of Commercial Analog Computers", *Computers and Automation*, Vol. 10, June 1961, pp. 117–118

Anon., " Survey of Commercial Analog Computers", *Computers and Automation*, Vol. 10, June 1962, pp. 130–132

Anon., "The Aircraft Industry Display: Computers" *Engineering*, Vol, 180, 30 Sept 1955, pp. 471–474.

Anon., "The Boeing Electronic Analog Computer: The use of analog equipment in the study of dynamic systems", *Internal Typescript Report*, Physical Research Unit, Boeing Airplane Co., Seattle, c. 1950.

Anon., "Three-Dimensional Analogue Computer: Predicting High-speed Flight", *Engineering*, 5 Nov 1954, pp. 596–597.

Anon., "Three-Dimensional Analogue Computer", *The Engineer*, 15 Oct 1954, pp. 532–533.

Anon., *An Intercept System Using All-Weather Rocket-Armed Interceptors in the Period 1956–1960*. Vol. 1 of 2, 1 March 1955. AT&T Archives, Case 26656–100.

Anon., *Automatic Carrier Landing Study: Final Report*, April 1954. AT&T Archives, Case 26656–89.

Anon., *Computing Machine Service*, Bureau of Ships, Department of the Navy, Navyships 250–223–1, 1 June 1953.

Anon., *Electronic Analogue Computation Laboratory: Bulletin No. 2*, Department of Engineering, UCLA, Ca., 12 Dec 1950.

Anon., *Instruments & Automation*, Vol. 30, Nov 1957, p. 2089.

Anon., *Instruments & Control Systems*, Vol. 34, Aug 1961, p. 1478.

Anon., *Instruments & Control Systems*, Vol. 34, April, 1961 pp. 681–682.

Anon., *Instruments & Control Systems*, Vol. 37, Nov 1964, p. 144.

Anon., *Interceptor Terminal Positions and Headings from Simulated Vectoring Operations*, 29 Feb 1956, AT&T Archives, Case 26656–100.

Anon., *Navy Intercept Project Quarterly Engineering Report No. 18*, 1 Oct 1953 to 1 Jan 1954.

Anon., *Navy Intercept Project Quarterly Engineering Report No. 18*, 1 Oct 1953–1 Jan 1954. AT&T Archives, Case 26656–100.

Anon., *The Simulation Council News Letter*, in *Instruments & Automation*, Vol. 31, Feb 1955.

Anon., *The Simulation Councils News Letter*, in *Instruments & Control systems*, Vol. 32, May 1959.

Anon., *Simulation Council News Letter*, in *Instruments & Automation*, Vol. 29, Feb 1956, pp. 301–304.

Anon., *Simulation Councils News Letter*, in *Instruments & Automation*, Vol. 29, Aug 1956, p. 1567.

Anon., *Simulation Council News Letter*, The Simulation Council, Camarillo, Ca., May 1953, pp. 4–7.

Argyris, J. H., "The Impact of Digital Computers on Engineering Science," *The Aeronautical Journal of the Royal Aeronautical Society*, Vol. 74, Feb 1970, pp. 111–127.

Armer, P., Smithsonian Computer History Project—Oral History Collection, National Museum of American History Archive Centre, Washington DC., Typescript of Interview, 17 April 1973.

Arthur, W. B., "Competing technologies: an overview", in G. Dosi, C. Freeman, R. Nelson, G. Silverburg and L. Soete (eds), *Technical Change and Economic Theory*, Pinter, London, 1988, pp. 590–607.

Aspray, W. "Edwin L. Harder and the Anacom: Analogue Computing at Westinghouse", *Annals of the History of Computing*, Vol. 15 no. 2, 1993, pp. 35–52.

Aspray, W., "The History of Computing Within the History of Information Technology", *History and Technology*, Vol. 11, no. 1, 1994, pp. 7–19.

Aspray, W. (ed.), *Computing Before Computers*, Iowa State University Press, Ames, Iowa, 1990.

Aspray, W., *John von Neumann and the Origins of Modern Computing*, MIT Press, Cambridge, Mass., 1990.

Augarten, S., *Bit by Bit: An Illustrated History of Computers*, George Allen & Unwin, London, 1985.

Barnes, B. "The Science-Technology Relationship: A Model and A Query", *Social Studies of Science*, Vol. 12, 1982, pp. 166–172.

Barnett, R. M., "NASA Ames Hybrid Computer Facilities and their Application to Problems in Aeronautics", *International Symposium on Analogue and Digital Techniques Applied to Aeronautics*, Liege, Belgium, Sept 1963.

Barney, G. C., and J. N. Hambury, "Developing the UMIST system", *Computer Weekly*, 13 Aug 1970, p. 10.

Bauer, W. F., "Aspects of Real-Time Simulation", *Trans., IRE, Electronic Computers*, EC-7, 1958, pp. 134–136.

Bauer, W. F., and G. P. West, "A System for General-Purpose Analog–Digital Computation", *Journal of the Association for Computing Machinery*, Vol. 4, Jan 1957, pp. 12–17.

Baxter, J. P., *Scientists Against Time*, Little, Brown, Boston, Mass., 1946.

Beard, R. E., "The Construction of a Small-Scale Differential Analyzer and its Application to the Calculation of Actuarial Functions", *Journal of the Institute of Actuaries*, Vol. 71, 1941, pp. 193–227.

Bekey, G. A., and W. J. Karplus, *Hybrid Computation*, John Wiley and Sons, New York, 1968.

Bell Telephone Laboratories, *A History of Engineering and Science in the Bell System: National Service in War and Peace (1925–1975)*, Bell Telephone Laboratories, 1978.

Bennett, W. C., "The Interconnection of Remotely Located Analog Computer Equipment", *The Simulation Council Newsletter*, in *Instruments & Control Systems*, Vol. 34, Aug 1961, pp. 1475–1486.

Bijker, W. E., T. P. Hughes and T. Pinch, (eds.), *The Social Construction of Technological Systems: New Directions in the Sociology and History of Technology*, MIT Press, Cambridge, Mass., 1987.

Black, H. S., "Stabilised Feedback Amplifiers", *Bell System Technical Journal*, Vol. 13, Jan 1934, pp. 1–18.

Blackett, P. M. S., and F. C. Williams, "An Automatic Curve Follower for Use With the Differential Analyzer", *Proc., Cambridge Philosophical Society*, Vol. 35, 1939, pp. 494–505.

Blake, R. G., P. M. Pigott, R. J. Smale, R. K. W. Bowdler and M. J. Whitmarsh-Everises, "Hybrid Computer Simulation of Dungenesss 'B' Power Station", *Engineering Division Report, No. RD/C/R52*, CEGB, London, May 1967.

Blanyer, C. G., and H. Mori, "Analogue, Digital and combined Analogue-Digital Computers for Real-Time Simulation, *Proc., Eastern Joint Computer Conference*, Washington DC, Dec 1957, pp. 104–110.

Bode, H. W., *Network Analysis and Feedback Amplifier Design*, Van Nostrand, New York, 1945.

Boltz, C. L., "Business with a big future", *Computer Weekly*, Hybrids and Analogues Supplement, 11 Jan 1968, p. 1.

Boothroyd, A. R., E. C. Cherry and R. Makar, "An electrolytic tank for the measurement of steady-state response, transient response, and allied properties of networks", *Proc., IEE*, Vol. 96, Pt 1, 1949, pp. 163–177.

Bower, T., *The Paperclip Conspiracy: The Battle for the Spoils and Secrets of Nazi Germany*, Michael Joseph, London, 1987.

Bratt, J. B., J. E. Lennard-Jones and M. V. Wilkes, "The Design of a Small Differential Analyzer", *Proc. Cambridge Philosophical Society*, Vol. 35, 1939, pp. 485–493.

Braun, H-J. "Introduction [to Symposium on 'Failed Innovations']", *Social Studies of Science*, Vol. 22, no. 2, May 1992, pp. 213–230.

Braun, H-J. "The Chrysler Automotive Gas Turbine Engine, 1950–80, *Social Studies of Science*, Vol. 22, no. 2, May 1992, pp. 339–352.

Breedon, D. B., "Analog versus digital techniques for engineering problems", *Trans., IRE*, IE-4, 1957, pp. 86–89.

Brennan, D., and H. Sano, "Pactolus—a digital analog simulator program for the IBM 1620", *Proc., Fall Joint Computer Conference*, 1964, pp. 299–312

Bromley, A. G. "Analogue Computing Devices", pp. 156–199, in W. Aspray, (ed.), *Computing Before Computers*, Iowa State University Press, Ames, Iowa, 1990.

Brown, E., and P. M. Walker, "Stardac—A General Purpose Analogue Computer: Part II", *Applied Electronics Laboratory Report No. SL.202*, General Electric Company, *Applied Electronics Laboratory*, 1 April 1957.

Brown, F. A., "Hybrid Simulation Techniques", *Simulation Council News Letter*, in *Instruments & Control Systems*, Vol. 36, Sept 1963, pp. 145–160.

Brown, G. S., quoted in, H. P. Peters, "AC Network Calculator Market Survey", *Typescript report*, Georgia Institute of Technology, Atlanta, Georgia, 1954, pp. 1–8, pp. 5–6.

Brown, M., "Which System to Choose?", *Computer Weekly*, 11 Jan 1968, p. 5.

Brown, M. G., "An Introduction to Hybrid Computing", *Typescript report*, British Aircraft Corp., Stevenage, 1976.

Bubloz, R., "Appreciating the analogue and hybrid techniques", *Computer Weekly*, 30 Aug 1973, pp. 6, 12.

Buchanan, R. A., "The Atmospheric Railway of I. K. Brunel", *Social Studies of Science*, Vol. 22, no. 2, May 1992, pp. 231–244.

Buchanan, R. A., *The Engineers: A History of the Engineering Profession in Britain, 1750–1914*, Jessica Kingsley, London, 1989.

Burns, A. J., and R. E. Kopp, "Combined Analogue-Digital Computer Simulation", *Proc., Eastern Joint Computer Conference*, Washington DC, Dec 1961, pp. 114–123.

Bush, V., "Instrumental Analysis", *Bulletin of the American Mathematical Society*, Vol. 42, 1936, pp. 649–669.

Bush, V., and H. Hazen, "Integraph Solutions of Differential Equations", *Journal of the Franklin Institute*, No. 204, 1927, pp. 575–615.

Bush, V., and H. Hazen, "The Differential Analyzer: A New Machine for Solving Differential Equations", *Journal of the Franklin Institute*, Vol. 212, 1931, pp. 447–488.

Bush, V., and S. Caldwell, "A New Type of Differential Analyzer", *Journal of the Franklin Institute*, Vol. 240, 1945, pp. 255–326.

Bush, V., F. D. Gage and H. R. Stewart, "A Continuous Integraph," *Journal of the Franklin Institute*, Vol. 203, 1927, pp. 63–84.

Bush, V., *Pieces of the Action*, William Morrow, New York, 1970.

Cahn, L., quoted in J. McLeod, "Ten Years of Computer Simulation", *Trans., IRE, Electronic Computers*, EC-11, Feb 1962, pp. 2–6.

Callon, M., " Mapping the Dynamics of Science and Technology: The Case of the Electric Vehicle", in: Callon, M., J. Law and A. Rip (eds) *Mapping the Dynamics of Science and Technology: Sociology of Science in the Real World*, Macmillan Press, London, 1986.

Campbell-Kelly, M., *ICL: A Business and Technical History*, Clarendon, Oxford, 1989.

Campbell-Kelly, M and W. Aspray, *Computer: A History of the Information Machine*, Basic Books, New York, 1996.

Carroll, J. M., "Analog Computers for the Engineer", *Electronics*, June 1956, pp. 122–129.

Cartmell, I. N. and R. W. Williams, "Guided Weapons Simulators", *Journal of the Royal Aeronautical Society*, Vol. 72, April 1968, pp. 356–360.

Cartwright, D., "Growth of the technique", *Computer Weekly*, 13 Aug 1970, pp. 9–12.

Ceruzzi, P. E., "Electronics Technology and Computer Science, 1940–1975: A Coevolution", *Annals of the History of Computing*, Vol. 10 no. 4, 1989, pp. 257–275.

Ceruzzi, P. E., *Beyond the Limits: Flight Enters the Computer Age*, MIT Press, Cambridge, Mass., 1989.

Ceruzzi, P. E., *Reckoners: The Prehistory of the Digital Computer, From Relays to the Stored Programme Concept, 1935–1945*, Greenwood Press, Westport, Conn., 1982.

Clancy, J., "Notes on the 'Bandwidth' of Digital Simulation (Technical Comment), *Simulation*, June 1967, pp. 19–20.

Clancy, J. L., and M. S. Fineberg, "Hybrid Computing: a users's view", *Simulation*, Aug 1965, pp. 104–112.

Clymer, A. B., "The Mechanical Analogue Computers of Hannibal Ford and William Newell", *Annals of the History of Computing*, Vol. 15 no. 2, 1993, pp. 19–34.

Cole, G. C., and S. L. H. Clarke, "The development of ARCH—a hybrid analogue digital system of computers for industrial control", *Journal of British IRE*, Vol. 26, 1963, pp. 45–58.

Collins, H., "Stages in the Empirical Programme of Relativism", *Social Studies of Science*, Vol. 11, 1981, pp. 3–10.

Computer Consultants, *The European Computer Users Handbook*, 5th Edition, Computer Consultants Ltd., Enfield, 1967.

Constant, E. W., "Scientific Theory and Technological Testability: Science, Dynamometers, and Water Turbines in the 19th Century", *Technology and Culture*, Vol. 24, No. 2, 1983, pp. 183–198.

Constant II, E. W.. *The Origins of the Turbojet Revolution*, Johns Hopkins University Press, Baltimore, 1980.

Conway, H. G., "Makers of Analogue Computers," *Engineering*, 23 Aug 1963, p. 231.

Coriolis, G. de, *Journal de Liouville*, Vol. 1, 1836.

Cortada, J., *A Bibliographical Guide to the History of Computing, Computers, and the Information Industry*, Greenwood Press, Westport, Conn., 1990.

Cowan, R. S., *More Work for Mother*, Basic Books, New York, NY, 1983.

Crank, J., *The Differential Analyzer*, Longmans, London, 1947.

Croarken, M., *The Centralisation of Scientific Computation in Britain: 1925-1955*, PhD Thesis, University of Warwick, 1985; published as, Croarken, M., *Early Scientific Computing in Britain*, Clarendon Press, London, 1990.

Crumb, Jr., C. B., "Engineering Uses of Analog Computing Machines", *Mechanical Engineering*, Vol. 74, no. 8, 1952, pp. 635–639.

Currie, A. A., "The General Purpose Analog Computer", *Bell Laboratories Record*, Vol. 29, no. 3, March 1951, pp. 101–107.

Daniels, G. H., (ed.), *Nineteenth Century American Science: A Reappraisal*, Northwestern University Press, Evanston, Ill., 1972.

Darker, H. A., "Analogue computing: equipment and performance", *The English Electric Journal*, Vol. 18, No. 5, Sept-Oct 1963, pp. 31–39.

Darwin, C. G., "Douglas Rayner Hartree: 1897–1958", *Biographical Memoirs of Fellows of the Royal Society*, Vol. 4, 1958, pp. 103–116.

de Florez, L., "Synthetic Aircraft", *Aeronautical Engineering Review*, April 1949, pp. 26–29.

Debroux, A., C. H. Green and H. D'Hoop, "Apache—A Breakthrough in Analogue Computing", *Trans., IRE, Electronic Computers*, EC-11, Oct 1962, pp. 699–706.

Denmead, J. K., J. S. Gatehouse, W. G. Littlejohn, M. W. Sage and G. W. T. White, "Analogue programming by digital computer" in *Proc., 5th AICA Conference, Lausanne*, 1967, Presses Academiques Europeennes, Brussels, 1967.

Dewey, B., (Vice President Reeves Inst.,) Letter to Professor William G. Shepard (University of Minnesota Institute of Technology), Minneapolis, 14 June 1949. CBI Archives, File: Marvin Stien 3–29.

Dian Laboratories Inc., Catalogue, Dian Laboratories Inc., New York, 1956.

Dickson, D., *The New Politics of Science*, Pantheon, New York, 1984.

Donner Model 30 Analogue Computer, Catalogue, Donner Scientific Co., 1955.

Donner Model 3000 Analogue Computer, Catalogue, Donner Scientific Co., 1957.

Dorfman, N., *Innovation and Market Structure: Lessons from the Computer and Semiconductor Industries*, Ballinger Books, Cambridge, Mass., 1987.

Dosi, G. "Technological Paradigms and Technological Trajectories: A Suggested Interpretation of the Determinants of Technical Change", *Research Policy*, Vol. 11, 1982, pp. 147–162.

Dummer, G. W. A., "Aids to Training—The Design of Radar Synthetic Training Devices for the RAF," *IEE Journal*, Vol. 96, Pt. III, 1949, pp. 101–116.

EAI 231R-V Information Manual, Electronic Associates Inc., Long Branch, NJ, 1964.

EAI Annual Report, Electronic Associates Inc, Long Branch, NJ, 1960.

EAI Annual Report 1960, Electronic Associates Inc, Long Branch, NJ, 1961.

EAI Annual Report, Electronic Associates Inc, Long Branch, NJ, 1962.

EAI Annual Report, Electronic Associates Inc, Long Branch, NJ, 1968.

EAI Annual Report, Electronic Associates Inc, Long Branch, NJ, 1970.

EAI Annual Report, Electronic Associates Inc, Long Branch, NJ, 1975.

EAI PACE User Group Survey, Typescript report, c. 1962;

Eames, C., and R. Eames, *A Computer Perspective*, Harvard University Press, Mass., 1973.

EASE Computer, Berkeley Division, Beckman Instruments Inc., c. 1954.

Edgerton, D., "Tilting at Paper Tigers", *British Journal for the History of Science*, Vol. 26, no. 88, March 1993, pp. 67–75.

Edgerton, D., *England and the Aeroplane: An Essay on a Militant and Technological Nation*, Macmillan, London, 1991.

Edwards, A. G., "It began with computers", in J. McLeod, (ed.), *Pioneers and Peers*, The Society for Computer Simulation International, San Diego, Ca., 1988, pp. 62–65.

Edwards, A. G., et al., *Advanced Hybrid Computer Systems*, US Army Materiel Command, Alexandria, Va., June 1973.

Edwards, P. N., *The Closed World: Computers and the Politics of Discourse in Cold War America*, MIT Press, Massachusetts, 1996.

Efmertová, M., "Czech Physicist Jaroslav Safránek and his Television", *Social Studies of Science*, Vol. 22, no. 2, May 1992, pp. 283–300.

Elzen, B., and D. Mackenzie, "From Megaflops to Total Solutions", in W. Aspray, (ed.), *Technological Competitiveness: Contemporary and Historical Perspectives on the Electrical, Electronics, and Computer Industries*, IEEE Press, Piscataway, NJ, 1993.

Emme, E. M., *Aeronautics and Astronautics*, National Aeronautics and Space Administration, Washington DC, 1961.

Enslow Jr. (Lt. Col.), P. H., "Hybrid Computer Technology in Europe: A State of the Art Survey", *ERO Technical Report No. ERO-1–74*, US Army European Research Office, London, 15 Feb 1974.

Evans, C., *The Making of the Micro: A History of the Computer*, Victor Gollancz, London, 1981.

Facilities and Services of Redifon-Astrodata Ltd, RAL, Littlehampton, 1966.

Fahrney, D. S., (R. Admiral), "Guided Missiles: US Navy The Pioneer", *American Aviation Society Journal*, Vol. 27, 1982, pp. 15–28.

Fairey Multi-purpose Analogue Computers, Trade Publication, Fairey Aviation, Heston, c. 1958.

Ferguson, E. S., "The Mind's Eye: Nonverbal Thought in Technology", *Science*, Vol. 197, 1977, pp. 827–836.

Ferguson, E. S., "Towards a discipline of the history of technology" *Technology and Culture*, Vol. 15, no. 1, 1974, pp. 13–24.

Ferguson, E. S., *Engineering and the Mind's Eye*, MIT Press, Cambridge, Mass, 1992.

Fifer, S., "Introduction" *Project Cyclone Symposium 1 on REAC Techniques*, Reeves Instrument Corporation Under Contract With Special Devices Centre, New York, March 15–16, 1951.

Fifer, S., *Analogue Computation: Theory, Techniques and Applications*, 4 Vols., McGraw-Hill, New York, 1961.

Flamm, K., *Targeting the Computer: Government Support and International Competition*, The Brookings Institution, Washington, DC, 1987

Flamm, K., *Creating the Computer: Government Industry and High Technology*, Brookings Institution, Washington DC, 1988.

Fogarty, L. E., and R. M. Howe, "Flight Simulation of Orbital and Reentry Vehicles", *Trans., IRE, Electronic Computers, EC-11*, Aug 1962, pp. 36–42.

Freeman, C., *The Economics of Industrial Innovation*, Frances Pinter, London, 1982.

Friedel, R., "Parkeside and Celluloid: The Failure and Success of the First Modern Plastic", *History and Technology*, Vol. 4, 1979, pp. 45–62.

Friedel, R. (ed.), Special Issue on "Engineering in the Twentieth Century", *Technology and Culture*, Vol. 27, no. 4, 1986.

Frost, S., "Compact Analog Computers", *Electronics*, July 1948, pp. 116–20.

Gait, J. J., "Analogue Computers and their Applications", *Process Control and Automation*, Nov 1958, pp. 485–480.

Gait, J. J., "Tridac: A Large Analogue Computer for Flight Simulation", *Journee International de Calcul Analogique*, Sept 1955.

Gait, J. J. and D.W. Allen, "An Electronic Analyser for Linear Differential Equations", *RAE Report: Guided Weapons 1*, December 1947.

Gardner, C., *British Aircraft Corporation: A History*, Batsford, London, 1981.

GEDA Price List, Goodyear Aircraft Corporation, Akron, Ohio, 1 March 1954.

GEDA: Analog-computing equipment, Trade publication, Goodyear Aircraft Corp. Akron, Ohio, 1954.

Goldberg, E. A., "Stabilization of Wide-Band Direct-Current Amplifiers for Zero and Gain", *RCA Review*, June 1950, pp. 296–300.

Goldberg, A. (ed.), *A History of Personal Workstations*, Addison-Wesley, Reading, Mass., 1988.

Goldstine, H. H., *The Computer: from Pascal to von Neumann*, Princeton University Press, Princeton, 1972.

Gomperts, R. J., and D. W. Righton, "LACE: The Luton Analogue Computing Engine, Pt 1", *Electronic Engineering*, Vol: 29, July 1957, pp. 306–312.

Gomperts, R. J., Computing applications where analogue methods appear to be superior to digital", *Journal Brit., IRE*, Aug 1957, pp. 421–428.

Gray, E. P., and J. W. Follin, Jr., "First Report on the REAC Computer", *Report TG-50*, The Applied Physics Laboratory, Johns Hopkins University, 15 Dec 1948.

Greenstein, J. L., "Application of ADDAVerter System in Combined Analogue–Digital Computer Operations", Paper Presented at *AIEE Pacific General Meeting*, Paper No. 56–842, June 1956.

Gunning, J. E. P., "The Rocket Propulsion Establishment, Westcott," *Journal of the Royal Aeronautical Society*, Vol. 70, Jan 1966, pp. 285–287.

Gunning, W. F., "A Survey of Automatic Computers, Analogue and Digital", *Report P-356*, The RAND Corporation, Santa Monica, Ca., 23 Dec 1953.

Gunning, W. F., Smithsonian Computer History Project—Oral History Collection, National Museum of American History Archive Centre, Washington DC., Typescript of Interview, 9 Oct 1972.

Hacker, B. C., "The Gemini Paraglider: A Failure of Scheduled Innovation, 1961–64", *Social Studies of Science*, Vol. 22, no. 2, May 1992, pp. 387–406.

Hall, A. C., "A Generalized Analogue Computer for Flight Simulation", *Trans., AIEE*, Vol. 69, 1950, pp. 308–312.

Harder, E. L., and G. D. Cann, "A Large-Scale General-Purpose Electric Analogue Computer", *Trans., AIEE*, Vol. 67, 1948, pp. 664–673.

Hare, R. E., and W. E. Willison, "Analogue Computers: Part 1", *Instrument Practice*, Nov 1958, pp. 1200–1207.

Hare, R. E., and W. E. Willison, "Analogue Computers: Part 2", *Instrument Practice*, Dec 1958, pp. 1295–1303.

Harris, H. E., "New Techniques for Analogue Computation," *Instruments & Automation*, Vol 30, May 1957, pp. 894–899.

Hartree, D. R., "A Great Calculating Machine", *Proc., Royal Institution*, Vol. 31, 1940, pp. 151–170.

Hartree, D. R., "Approximate Wave Functions and Atomic Field of Mercury", *Physical Review*, Vol. 46, 1934, pp. 738–743.

Hartree, D. R. "The Bush Differential Analyzer and its Applications", *Nature*, Vol. 146, 7 Sept 1940, pp. 319–323.

Hartree, D. R., "The Differential Analyzer", *Nature*, Vol. 135, 1935, pp. 940–943.

Hartree, D. R., "The ENIAC: An Electronic Computing Machine", *Nature*, Vol. 158, 1946, pp. 500–506.

Hartree, D. R. "The Mechanical Integration of Differential Equations", *Mathematical Gazette*, Vol. 22, 1938, pp. 342–364.

Hartree, D. R., and A. Porter, "The Construction and Operation of a Model Differential Analyzer", *Memoirs Manchester Lit. and Phil. Society*, Vol. 79, no. 5, 1935, pp. 51–73.

Hartree, D. R., and A. K. Nuttall, "The Differential Analyzer and its Applications in Electrical Engineering", *Journal of the Institute of Electrical Engineers*, Vol. 83, 1938, p. 648.

Hartree, D. R. and J. R. Womersley, "A Method for the Numerical or Mechanical Solution of Certain Types of Partial Differential Equations", *Proc. Royal Society*, Pt. A, Vol. 161, 1937, pp. 353–366.

Hartree, D. R., *Calculating Machines: Recent and Prospective Developments and their impact on Mathematical Physics*, Cambridge University Press, Cambridge, 1947.

Hartsfield, E., "Timing Considerations in a Combined Simulation System Employing a Serial Digital Computer", *Proc., Combined Analogue-Digital Computer Systems Symposium*, Philadelphia, Pa, Dec 1960.

Hass, A., "Gigantic Computer Goes to Junk Pile", *Philadelphia Inquirer*, Philadelphia, Pa., 19 Sept 1968.

Hassig, L. (ed.), *Alternative Computers*, part of the *Understanding Computers Series*, Time-Life Books, Alexandria, Virginia, 1989.

Hayward, K., *The British Aircraft Industry*, Manchester University Press, Manchester, 1989.

Heaviside, O., *Electrical Papers* Macmillan, New York, 1892.

Hendry, J., *Innovating for Failure*, MIT Press, Cambridge, Mass., 1989.

Hermann, P., *The Simulation Council Newsletter*, in *Instruments & Automation*, Vol. 28, Aug 1955, pp. 1307–1311.

Hermann, P. J., K. H. Starks and J. A. Rudolf "Basic Applications of Analog Computers", *Instruments & Automation*, Vol. 29, March 1956, pp. 464–469.

Higgins, W. H., B. D. Holbrook and J. W. Emling, "Defence Research at Bell Laboratories: Electrical Computers for Fire Control.", *Annals of the History of Computing*, Vol. 4, No. 3, July, 1982, pp. 218–236.

Hodges, A., *Alan Turing: The Enigma of Intelligence*, Burnett Books and Hutchinson, London, 1983.

Hodges, P. L., "Stardac—A General Purpose Analogue Computer: Part I", *Report No. SL.196*, General Electric Company, Applied Electronics Laboratory, 1 Feb 1957.

Hoffman, Q. J., H. B. Sabin and H. J. Smith, "Requirements for Computing Facilities in the Sperry Engineering Division", *Internal report 5281-7142*, Sperry Gyroscope Co., Great Neck, 15 Jan 1953, pp. 1–26, Hagley Museum and Library, Sperry Collection.

Holley Jr., I. B., "A Detroit dream of mass-produced fighter aircraft: The XP-75 fiasco", *Technology and Culture*, Vol. 28, no. 3, 1987, pp. 578–593.

Holst, P. A., "George A. Philbrick and Polyphemus—The First Electronic Training Simulator", *Annals of the History of Computing*, Vol. 4, No. 3, April 1982, pp. 143–156.

Holst, P. A., "Sam Giser and the GPS Instrument Company: Pioneering Compressed-time Scale (High-speed) Analog Computation", in J. McLeod (ed.), *Pioneers and Peers*, Society for Computer Simulation International, San Diego, Ca., 1988, pp. 4–10.

Hosenthien, H. H., and J. Boehm, "Flight Simulation of Rockets and Spacecraft", in E. Stuhlinger, F. I. Ordway, III, J. C. McCall and G. C. Bucher, (eds.), *From Peenemunde to Outer Space*, NASA George C. Marshal Space Flight Centre, Huntsville, 1962, pp. 437–469.

Hovious, R. L., C. D. Morrill and N. P. Tomlinson, "Industrial Uses of Analog Computers", *Instruments and Automation*, Vol. 28, April 1955, pp. 594–601.

Hughes, T. P., "Technological Momentum in History: Hydrogenation in Germany, 1898–1933", *Past and Present*, no. 44, Aug 1969, pp. 106–132.

Hughes, T. P., "The Evolution of Large Technological Systems", in Bijker, W. E., T. P. Hughes and T. Pinch, (eds.), *The Social Construction of Technological Systems: New Directions in the Sociology and History of Technology*, MIT Press, Cambridge, Mass., 1987.

Hughes, T. P., *Networks of Power, Electrification in Western Society, 1880–1930*, Johns Hopkins University Press, Baltimore, 1983.

Hult, J., "The Itera Plastic Bicycle", *Social Studies of Science*, Vol. 22, no. 2, May 1992, pp. 373–386.

Hunter, L. C., "The living past in the Appalachias of Europe: Watermills in southern Europe", *Technology and Culture*, Vol. 8, no. 4, 1967, pp. 446–466.

Ingerson, W. E., "The Nike-Ajax computer", *Bell Laboratories Record*, Vol. 38, no. 1, 1960, pp. 26–30.

Inston, H. H., "The hybrid data processing system at Blacknest", Typescript report, UK Atomic Energy Authority, Blackness, 26 Sept 1963, pp 1–16.

Inston, H. H., *et al.*, "A description of the AWRE hybrid data processing system, with some applications", *Proc., Association Internationale pour le Calcul Analogique, 4th Int., Conf.*, Brighton, Sept 1964.

Jackson, A. S., *Analog Computation*, McGraw-Hill, New York, 1960.

Jacobs Jr., H., "Equipment Reliability as Applied to Analogue Computers," *Journal of the Association for Computing Machinery*, Vol. 1, no. 1, Jan 1954, pp. 21–26.

Jacobsen, C. A., "The Application of the Hybrid Computer in Flight Simulation", *Proc., IBM Scientific Computing Symposium on Computer-Aided Experimentation*, 1966, pp. 206–244.

Jeffries, T. O., C. P. Newport, H. A. Darker and R. A. Flint, "The Saturn Analogue Computer", *Nuclear Power*, Jan 1962, pp. 69–72.

Jeffries, T. O., C. P. Newport, H. A. Darker and R. A. Flint, "A Large Analogue Computing Installation", *Nuclear Power*, Dec 1961, pp. 80–84.

Jenners, R. R., *Analog Computation and Simulation: Laboratory Approach*, Allyn and Bacon, Boston, 1965.

Johnson, C. L., *Analog Computer Techniques*, 2ed., McGraw-Hill, New York, 1963.

Johnson, S. O., and J. N. Grace. "Analog Computation in Nuclear Engineering", *Nucleonics*, Vol. 15, no. 5, May 1957, pp. 72–75.

Jones, H. M., "Ideas, history, technology", *Technology and Culture*, Vol. 1, no. 1, 1959, pp. 20–27.

Jones, J. C., and D. Readshaw, "LACE: The Luton Analogue Computing Engine: Pt 2", *Electronic Engineering*, Vol. 29, July 1957, pp. 306–312.

Karen, A., and B. Loveman, "Large-Problem Solutions at Project Cyclone", *Instruments and Automation*, Vol. 29, Jan 1956, pp. 78–83.

Kayan, C. F., "An Electrical Geometrical Analogue for Complex Heat Flow", *Trans., ASME*, Vol. 67, 1945, pp. 713–718.

Kidder, J. T., *The Soul of a New Machine*, Atlantic-Brown, Boston, Mass., 1981.

Kirchhoff, G., *Annals of Physics, Leipzig*, Vol. 64, 1845, p. 497.

Kirchhoff, G., *Collected Papers*, Leipzig, 1882.

Kline, R., "Science and Engineering Theory in the Invention and Development of the Induction Motor, 1880–1900", *Technology and Culture*, Vol. 28, No. 2, 1987, pp. 283–313.

Knight, A. J., "A Survey of Computing Facilities in the UK" (2nd ed.), *Directorate of Weapons Research Report No. 5/56*, Ministry of Supply, London, August 1956.

Korn, G. A., "Progress of Analogue/Hybrid Computation", *Proc., IEEE*, Vol. 54, no. 12, Dec 1966, pp, 1835–1848.

Korn, G. A., "Progress of Analogue/Hybrid Computation", *Proc., Fifth International Analogue Computation Meeting*", Lausanne, 28 Aug–2 Sept 1967, pp. 22–34.

Korn, G. A., "The Impact of the Hybrid Analog-Digital Techniques on the Analog Computer Art", *Proc. IRE*, Vol. 50, No. 5, 1962, pp. 1077–1086.

Korn, G. A., and T. M. Korn, *Electronic Analogue and Hybrid Computers*, 2ed., McGraw-Hill, New York, 1972.

Korn, G. A., and T. M. Korn, *Electronic Analogue Computers (D-C Analog Computers)*, McGraw-Hill, New York, 1952.

Korn, G. A., and T. M. Korn, *Electronic Analogue Computers (D-C Analog Computers)*, 2ed., McGraw-Hill, New York, 1956.

Kranakis, E., "Social Determinants of Engineering Practice: A Comparative View of France and America in the Nineteenth Century", *Social Studies of Science*, Vol. 19, no. 1, Feb 1989, pp. 5–70.

Kuehni, H. P., and H. A. Peterson, "A New Differential Analyzer", *Electrical Engineering*, Vol. 63, 1944, pp. 221–28.

Kuhn, T. S., *The Structure of Scientific Revolutions*, 2nd ed., Chicago University Press, Chicago, 1970.

Lakatos, E., "Interim Report on the Electrical Analog Computer", *Internal typescript Memorandum MM-46–10–60*, Bell Telephone Laboratories, 28 June 1946, pp. 1–26. AT&T Archives, Case 20878.

Lakatos, E., "Problem solving with the analog computer", *Bell Laboratories Record*, Vol. 29, no. 3, 1951, pp. 109–114.

Landauer, P., quoted in Hassig, L. (ed.), *Alternative Computers*, part of the *Understanding Computers Series*, Time-Life Books, Alexandria, Virginia, 1989.

Latour, B., *Science in Action*, Open University Press, Milton Keynes, 1986.

Laudan, R., (ed.), *The Nature of Technological Knowledge: Are Models of Scientific Change Relevant?*, D. Reidel, Dordrecht, 1984.

Lavington, S. H., *Early British Computers: The Story of Vintage Computers and the People Who Built Them*, Manchester University Press, Manchester, 1980.

Layton, Jr., E. T., "Mirror Image Twins: The Communities of Science and Technology in Nineteenth Century America", *Technology and Culture*, Vol. 12, no. 4, 1974, pp. 562–580.

Layton, Jr., E. T., "Scientific Technology, 1845–1900: The Hydraulic Turbine and the Origins of American Industrial Research", *Technology and Culture*, Vol. 20, no. 1, 1979, pp. 64–89.

Layton, Jr., E. T., "Technology as Knowledge", *Technology and Culture*, Vol. 15, no. 1, 1974, pp. 31–41.

Layton, Jr., E. T., "Through the Looking Glass, or News from Lake Mirror Image", *Technology and Culture*, Vol. 29, no. 3, 1987, pp. 594–607.

Layton, Jr., E. T., *The Revolt of the Engineers: Social Responsibility and the American Engineering Profession*, Johns Hopkins Univ. Press, Baltimore, 1986.

Leger, R. M., "Requirements for Simulation of Complex Systems", *Proc., Flight Simulation Symposium*, Sept 1957, pp. 125–131.

Leger, R. M., "Specification for Analogue–Digital–Analogue Converting Equipment for Simulation Use", Paper Presented at *AIEE Pacific General Meeting*, Paper No. 56–860, June 1956.

Lewis, W. D., and W. F. Trimble, "The Airmail Pickup System of All American Aviation: A Failed Innovation?", *Social Studies of Science*, Vol. 22, no. 2, May 1992, pp. 301–316.

Linebarger, R. N., and D. Brennan, "A Survey of Digital Simulation", *Simulation*, Vol 3, Dec 1964, pp. 22–36.

Lovell, C. A., "Continuous Electrical `Computation", *Bell Laboratories Record*, Vol. 25, no. 3, March 1947, pp. 114–118.

Lovell, C. A., D. B. Parkinson and B. T. Weber, US Patent, No. 2,404,387, filed May 1, 1941; issued July 23, 1946.

Lundstrom, D. E., *A Few Good Men From Univac*, MIT Press, Cambridge, Mass., 1987.

MacKay D. M., and M. E. Fisher, *Analogue Computing at Ultra-high Speeds*, John Wiley and Sons, New York, 1962.

Mackenzie, D., and J. Wajcman, (eds.), *The Social Shaping of Technology*, Open University Press, Milton Keynes, 1985.

Mackenzie, D., *Inventing Accuracy: A Historical Sociology of Nuclear Missile Guidance"*, MIT Press, Cambridge, Mass., 1990.

MacLusky, G. J. R., "An Analogue Computer for Nuclear Power Studies," *Proc., IEE*, Vol. 104, Pt. B, no. 17, Sept 1957, pp. 433–442.

Mahoney, M. C., "The history of computing in the history of technology", *Annals of the History of Computing*, Vol. 10, no. 2, 1988, pp. 113–125.

Makinson, W., and G. M. Hellings, "Synthetic Aids to Flying Training," *Journal of the Royal Aeronautical Society*, Vol, 61, Aug 1957, pp. 509–528.

Massey, H. S. W., J. Wylie, R. A. Buckingham and R. Sullivan, "Small Scale Differential Analyzer: its Construction and Operation", *Proc. Royal Irish Academy*, Pt. A, Vol. 45, 1938, pp. 1–21.

Maynall, J. D. "Electrical Analogue Computing", *Electronic Engineering*, Vol. 19, 1947: Part 1, June, pp. 178–180. Part 2, July, pp. 214–217, Part 3, August, pp. 259–262.

McCool, W. A., "Electronic Analog Computation", *Naval Research Laboratory Report*, Washington DC, December 1950.

Mccoy, R., *et al.* "Designing an Office-Size Electronic Analog Computer", *Electrical Manufacturing*, Vol. 47, 1951, pp. 94–9, & pp, 230–2, & p.234.

McCutcheon, R. T., "Science, Technology and the State in the Provision of Low-Income Accommodation: The Case of Industrialized House Building, 1955–77", *Social Studies of Science*, Vol. 22, no. 2, May 1992, pp. 353–372.

McLeod, J., (ed.), *Pioneers and Peers*, The Society for Computer Simulation International, San Diego, Ca., 1988, pp. 4–10.

McLeod, J., *Simulation Council News Letter*, in *Instruments & Automation*, Vol 29, Oct 1956, pp. 2015–2023.

McLeod, J. H., and R. M. Leger, "Combined Analog and Digital Systems -Why, When and How", *Instruments & Automation*, Vol 30, June 1957, pp. 1126–1130.

Medwick, P. A., "Douglas Hartree and Early Computation in Quantum Mechanics", *Annals of The History of Computing*, Vol. 10, no. 2, 1988, pp. 105–111.

Meiksins, P., "The 'Revolt of the Engineers' Reconsidered", *Technology and Culture*, Vol. 29, No. 2, 1988, pp. 219–246.

Meissinger, H., "Trends in Simulation", *Simulation Council News Letter*, in *Instruments & Control Systems*, Vol. 32, Sept 1959, p. 1389–1393

Melahn, W., "Modifications of the RAND REAC", *Typescript report*, RAND Corporation, Santa Monica, Ca., 26 Feb 1951.

Melahn, W. S., "Modification of the RAND REAC", *Project Cyclone Symposium 1 on REAC Techniques*, Reeves Instrument Corporation Under Contract With Special Devices Centre, New York, 15–16 March 1951, pp. 41–44.

Mengel, A. S., "New Form of Euler Equations for Optimum Flight", *Research Memorandum RM-307*, Project RAND, Santa Monica, Ca., 6 Jan 1950.

Mengel, A. S., "Optimum Trajectories", *Project Cyclone Symposium 1 on REAC Techniques*, Reeves Instrument Corporation Under Contract With Special Devices Centre, New York, March 15–16, 1951, pp. 7–13.

Mengel, A. S., "Principles of the REAC", *Research Memorandum RM-226*, Project RAND, Santa Monica, Ca., 19 Jan 1949.

Mengel, A. S., and W. S. Melahn, "RAND REAC Operators' Manual", *Report RM-525*, Project RAND, Santa Monica, Ca., 1 Dec 1950.

Michel, J. G. L., "The Mechanical Differential Analyzer: Recent Developments and Applications", *Journees Association International du Calcul Analogique*, Sept 1955.

Miller, D., "Faster than real time", *Simulation*, Nov 1977, pp. 117–118.

Millichamp, D., "Computers help power users", *Engineering*, 24 July 1970, p. 88.

Moody's Computer Industry Survey, Moody's Investors Services Inc., New York, Vol. 1, no. 1, 1965.

Moody's Industrial Manual, New York, 1954.

Morton, P., *Fire Across the Desert*, Australian Government Publishing Service Press, Canberra, 1989.

Mott, N. F., "John Lennard-Jones", *Biographical Memoirs of Fellows of the Royal Society*, Vol. 1, 1955, pp. 175–184.

Mowery, D. C., and N. Rosenberg, *Technology and the Pursuit of Economic Growth*, Cambridge University Press, Cambridge, 1989.

Murphy, B., "This Analogue Centre Serves European and British Industry," *Automatic Data Processing*, March 1959, pp. 1–4.

Murray, F. J., *Mathematical Machines, Volume II: Analogue Devices*, Columbia University Press, New York, 1961.

Mynall, D. J., "Electrical Analogue Computing—Part 3—Functional Transformation", *Electronic Engineering*, Vol. 19, Aug 1947, pp. 259–262.

Nash, S., (ed.), *A History of Scientific Computing*, Addison-Wesley, Reading, Mass., 1990.

Nelson, R. R., and S. G. Winter, *An Evolutionary Theory of Economic Change*, Belknap/Harvard University Press, Cambridge, Mass., 1982.

Nenadal, Z., and B. Mirtes, *Analogue and Hybrid Computers*, SNTL, (Prague) 1962, English translation Iliffe Books (London), 1968.

New Encyclopedia Britannica, University of Chicago, Chicago, Vol. 9, 1991, pp. 496–497.

Niemann, C. W., "Bethlehem Torque Amplifier", *American Machinist*, Vol. 66, 1927, pp. 895–897.

Noble, D., *America by Design: Science, Technology, and the Rise of Corporate Capitalism*, Oxford University Press, Oxford, 1977.

Noble, D. F., *Forces of Production: A Social History of Industrial Automation*, Victor Gollancz, New York, NY, 1984.

Nyce, J. M., (ed.), Special issue on analogue techniques, *Annals of the History of Computing*, Vol. 18, no. 4, 1996.

Nyquist, H., " Regeneration Theory" *Bell System Technical Journal*, Vol. 11, Jan 1932, pp. 126–147.

Olmstead, R. W., *Simulation Council News Letter*, in *Instruments & Control Systems*, Vol. 36, Oct 1963, p. 117

Owens, L. "Vannevar Bush and the Differential Analyzer: The Text and Context of an Early Computer", *Technology and Culture*, Vol. 27, no. 1, 1986, pp. 63–95.

PACE User Group, Minutes of 2nd Meeting, CEBG Headquarters, Friars House, London, 9 Oct 1961.

Paquette, G. A., "Progress of Hybrid Computation at United Aircraft Research Laboratories", *Proc., Fall Joint Computer Conference*, 1964, pp. 695–706.

Parton, K. C., and D. R. Roberts, "Principles and Applications of a Control System Simulator", *G. E. C. Journal*, Vol. 26, no. 4, 1959, pp. 3–12.

Paskman, M., and J. Heid, "The Combined Analogue–Digital Computer System", *Proc., Combined Analogue–Digital Computer Systems Symposium*", Philadelphia, Pa, Dec 1960.

Paul, R. J. A., "Hybrid methods for function generation", *Cranfield, College of Aeronautics Report*, No. 153, 1961.

Paul, R. J. A., and G. C. Rowley, "Hybrid computing techniques", *Proc. Inst of Mechanical Eng.*, Dec 1962, pp 16–25.

Paul, R. J. A., and M. E. Maxwell, "Digital and hybrid simulation techniques", *Control*, Vol. 3, Apr 1960, pp. 120–124.

Paul, R. J. A., "The SHORT Electronic Analogue Computer", *The Overseas Engineer*, Vol. 29, 1956, pp. 205–208.

Paul, R. J. A. and E. L. Thomas, "The Design and Applications of a General-Purpose Analogue Computer", *Journal of the British IRE*, Vol. 17, Jan 1957, pp. 49–73.

Paynter, H. M. "In Memoriam: George A. Philbrick (1913–1974)", *Trans., of the ASME, Journal of Dynamic Systems, Measurement and Control*, June 1975, pp. 213–215.

Paynter, H. M., (ed.), *A Palimpsest on the Electronic Art*, George A. Philbrick Researches Inc, Boston, Mass., 1955.

Peak, D. C., "A Survey of Requirements for a Present-Day and Fourth Generation Hybrid Computer Systems in the United States", paper presented at, the *Special Symposium on Advanced Hybrid Computing*, sponsored by US Army Materiel Command, *Summer Computer Simulation Conference*, San Francisco, Ca., 24 July 1975.

Penick, J. L., *et al.* (eds.), *The Politics of American Science, 1939 to the Present*, MIT Press, Cambridge, Mass., 1972.

Perkins, C. D., "Man and Military Space", *Journal of the Royal Aeronautical Society*, Vol. 67, no. 631, July 1963, pp. 397–412.

Perry, R. L., "The Atlas, Thor, Titan and Minuteman", in E. M. Emme, (ed.), *The History of Rocket Technology*, Wayne State University Press, Detroit, 1964, pp. 142–161.

Peters, H. P., "AC Network Calculator Market Survey", *Typescript report*, Georgia Institute of Technology, Atlanta, Georgia, 1954.

Peterson, H. A., and C. Concordia, "Analyzers for Use in Engineering and Scientific Problems", *General Electric Review*, Vol. 49, 1945, pp. 29–37.

Petroski, H., *To Engineer is Human: The Role of Failure in Successful Design*, Macmillan, London, 1985.

Philbrick, G. A., "Designing Industrial Controllers by Analog", *Electronics*, Vol. 21, June 1948, pp. 108–111.

Philbrick, G. A., "The Philbrick Automatic Control Analyzer: Program for Development and Exploitation", *Hand written memorandum, the Foxboro Co., RD Project 10530*, 1 Feb 1940, pp. 1–3. Archives of the National Museum of American History, Washington, DC, Box: Philbrick.

Philbrick, G. A., and C. E. Mason, "Automatic Control in the Presence of Process Lags," *Trans., ASME*, Vol., 62, May 1940, pp. 295–308.

Philbrick, G. A., and H. Paynter, *Applications Manual for Philbrick Octal Plug-in Computing Amplifiers*, George A. Philbrick Researches Inc, Boston, Mass., 1956.

Philbrick, G. A., and H. M. Paynter, "The Electronic Analog Computor as a Lab Tool," *Industrial Laboratories*, Vol. 3, no. 5, May 1952.

Philbrick, G. A., W. T. Stark and W. C. Schaffer, "Electronic Analog Studies for Turboprop Control Systems," *SAE Quarterly Trans.*, Vol. 2, no. 2, April 1948.

Pigott, P. M., and R. J. Smale, "Hybrid Computer Study Using the EAI 8900 at EAI PCC", Minutes of the 25th Meeting, 26th July 1967, *PUG Newsletter*, Vol. 1. no. 2, Sept 1967, p. 2.

Pinch, T. J. and W. E. Bijker, "The Social Construction of Facts and Artifacts: or How the Sociology of Science and the Sociology of Technology Might Benefit Each Other", *Social Studies of Science*, Vol. 14, 1984, pp. 399–441.

Post, R. C., "The Page Locomotive: Federal sponsorship of invention in mid-19th-century America", *Technology and Culture*, Vol. 13, no. 2, 1972, pp. 140–169.

Proctor, W. G., and M. F. Mitchell, "The PACE Scaling Routine for Mercury", *The Computer Journal*, Vol. 5, 1962–63, pp. 24–27.

Project Cyclone Symposium 1 on REAC Techniques, Reeves Instrument Corporation Under Contract With Special Devices Centre, New York, March 15–16, 1951.

Project Cyclone Symposium 2 on Simulation and Computing Techniques, Reeves Instrument Corporation Under Sponsorship of the US Navy Special Devices Centre and the US Navy Bureau of Aeronautics, New York, April 29th to May 2nd, 1952.

Project Typhoon Symposium 3, Simulation and Computing Techniques, sponsored by Bureau of Aeronautics and US Naval Air Development Centre, Johnsville, Pa., Oct 12–14, 1953.

Pugh, E. W., *Memories That Shaped an Industry*, MIT Press, Cambridge, Mass., 1984.

Quarmby, R. B., "Electronic Analogue Computing: Survey of Modern Techniques", *Wireless World*, March, 1954, pp. 113–118.

Radio Corporation of America, "News Information: New RCA Electronic Computer Aids US Air Defence; ...", *Internal Typescript report*, RCA, New York, 21 Nov 1950, pp. 1–5.

RAL, *RAL, Facilities and Services of Redifon-Astrodata Ltd.*, RAL, Littlehampton, 1966.

RADIC: Redifon Analogue Digital Computing System, Trade Publication, Redifon Ltd, Crawley, c. 1960.

Ragazzini, J. R., R. H. Randall and F. R. Russell, "Analysis of Problems in Dynamic by Electronic Circuits", *Simulation*, Sept 1964, pp. 54–65.

Ragazzini, J. R., R. H. Randall and F. R. Russell, "Analysis of Problems in Dynamic by Electronic Circuits", *Proc., IRE*, Vol. 35, May 1947, pp. 444–452.

RCA an historical perspective, RCA, Cherry Hill, NJ, 1978.

Redifon, Computer Department News Letter, No. 3, 1963, pp. 1–4.

Redmond, K. C., and T. M. Smith, *Project Whirlwind: The History of a Pioneer Computer*, Digital Press, Bedford, Mass., 1980.

Rees, M., "The Federal Computing Machine Program", *Science*, Vol. 112, Dec 1950, pp. 731–736.

Reichardt, O. A., M. W. Hoyt and W. Thad Lee, "The Parallel Digital Differential Analyzer and its Applications as a Hybrid Computing Systems Element," *Simulation*, Feb 1965, pp. 104–126.

Reiners, C. A., "A Survey of Computing Facilities in the UK", *Directorate of Weapons Research (Defence) Report No. 13/53*, Ministry of Supply, London, Sept 1956.

Ridenour, L. N., "The Role of the Computer", *Scientific American*, Vol. 187, 1952, pp. 116–130.

Rogers A. E., and T. W. Connelly, *Analog Computation in Engineering Design*, McGraw-Hill, New York, 1960.

Rogers, S., Smithsonian Computer History Project—Oral History Collection, National Museum of American History Archive Centre, Washington DC., Typescript of Interview, 9 Aug 1973.

Rose, H. E., "The Mechanical Differential Analyzer: Its Principles, Development, and Applications, *Proc., Institute of Mechanical Engineers*, Vol. 159, 1948, pp. 46–54.

Rowlands, B., and W. B. Boyd, *US Navy Bureau of Ordnance in World War II*, Bureau of Ordnance, Department of the Navy, US Government Printing Office, Washington, DC, 1953.

Rubinfien, D., "What Computers Promise", *Control Engineering*, Sept 1954, pp. 64–67.

Rurup, R., "Historians and modern technology: Reflections on the development and current problems of the history of technology", *Technology and Culture*, Vol. 15, no. 2, 1974, pp. 161–193.

Sapolsky, H. M., "The Origins of the Office of Naval Research", in D. M. Masterson, (ed.), *Naval History: The Sixth Symposium of the US Naval Academy*, Wilmington, De., Scholarly Resources, 1987.

Schatzberg, E., *Ideology and Technical Change: The Choice of Materials in American Aircraft Design Between the World Wars*, PhD thesis, University of Pennsylvania, 1990, published as, Schatzberg, E., *Wings of Wood, Wings of Metal: Culture and Technical Choice in American Airplane Materials, 1914–1945*, Princeton University Press, Princeton, 1999

Schatzberg, E., "Ideology and Technical Choice: The Decline of the Wooden Airplane in the United States, 1920–1945", *Technology and Culture*, Vol. 35, no. 1, 1994, pp. 34–69.

Schwiebert, E. G., *A History of the US Air Force Ballistic Missiles*, Frederick A. Praeger, New York, 1965.

Scott, A. J., "The aerospace-electronics industrial complex of Southern California: The formative years, 1940–1960", *Research Policy*, Vol. 20, 1991, pp. 439–456.

Seely, B., "Research, Engineering, and Science in American Engineering Colleges: 1900–1960", *Technology and Culture*, Vol. 34, no. 2, pp. 344–386.

Seifert, W. W., "Seminar for Committee on Machine Methods of Computation", 27 Oct 1953. *Typescript*, MIT Archives, MC6.

Seifert, W. W., "The Role of Computing Machines in the Analysis of Complex Systems", Paper presented at the International Analogy Computation Meeting, Brussels, Sept 26–Oct 1, 1955, pp. 1–19. *Typescript*, MIT Archives, MC6.

Seifert, W. W., and H. Jacobs, Jr., "Problems Encountered in the Operation of the M. I. T. Flight Simulator", 11 April 1952. *Typescript*, MIT Archives MC6.

Seifert, W. W., and H. Jacobs, Jr., "The Role of Simulators in Relation to Low-Altitude Multiple-Target Problems", Paper presented at the US Redstone Arsenal, 5 Jan 1953. *Typescript*, MIT Archives, MC6.

Serlin, O., *Simulation Council News Letter*, in *Instruments & Control Systems*, Vol. 36, Feb 1963, p. 142.

Shapio, S., and L. Lapides, "A Combined Analogue–Digital Computer for Simulation of Chemical Processes", *Proc., Combined Analogue–Digital Computer Systems Symposium"*, Philadelphia, Pa., Dec 1960.

Sheingold, D. H., "George A. Philbrick: Gentleman, innovator," *Electronic Design*, Vol. 3, 1 Feb 1975, pp. 7.

Sherman, P. M., "A Report on DC Electronic Analogue Computation and the Aeronautical Division Analogue Computer Facility", *Internal report 5234–3522*, Sperry Gyroscope Co., Great Neck, NJ, 25 Aug 1955, pp. 1–148, Hagley Museum and Library, Sperry Collection.

Sherman, P. M., "The Theory, Operation and Use of Philbrick High-Speed Analogue Computer", *Internal report 5234–3458*, Sperry Gyroscope Co., Great Neck, 2 Sept 1955, pp. 1–56, Hagley Museum and Library, Sperry Collection.

Shorts General Purpose Analogue Computor, Trade Publication, Short Bros and Harland, Belfast, c. 1955.

Siefert, W. W., "The Role of the M. I. T. Flight Simulator in the Field of Analogue Computation", *Typescript*, c. 1952, MIT Archives MC6.

Skramstad, H. K., "Combined Analog–Digital Techniques in Simulation", in F. L. Alt and M. Rubinoff (eds.), *Advances in Computers*, Academic Press, New York, Vol. 3, 1962, pp. 275–298.

Skramstad, H. K., A. A. Ernst and J. P. Nigro, "An Analogue–Digital Simulator for the Design and Improvement of Man–Machine Systems, *Proc., Eastern Joint Computer Conf.*, Washington DC, Dec 1957, pp. 90–96.

Small, J. S., "General-purpose Electronic Analog Computing: 1945–1965", *Annals of the History of Computing*, Vol. 15, 1993, pp. 8–18;

Small, J. S., "Engineering, Technology and Design: The Post-Second World War Develop-ment of Electronic Analogue Computers", *History and Technology*, Vol. 11, no. 1, 1994, pp. 33–48.

Small, J. S., "Electronic Analog Computer", in *Instruments of Science: an Historical Encyclopedia*, Garland Publishing, New York, 1998.

Smith, F., "The Electronic Simulator for the Solution of Flutter and Vibration Problems", *RAE Report: Structures*, No. 51, October 1949.

Smith, G. W. and R. C. Wood, *Principles of Analog Computation*, McGraw-Hill, New York, 1959.

Smith, M. R., (ed.), *Military Enterprise and Technological Change: Perspectives on the American Experience*, MIT Press, Cambridge, Mass., 1985.

Snyder, S., Smithsonian Computer History Project—Oral History Collection, National Museum of American History Archive Centre, Washington DC., Typescript of Interview, 16 July 1970.

Softky, S., and J. Jungerman, "Electrolytic tank measurements in three dimensions", *Review Scientific Instruments*, Vol. 23, 1952, pp. 306–307.

Soroka, W., *Analog Methods in Computation and Simulation*, McGraw-Hill, New York, 1954.

Sorondo, V. J., R. B. Wavell and F. E. Nixon, "Cost and Speed Comparisons of Hybrid VS. Digital Computers", paper presented at, the *Special Symposium on Advanced Hybrid Computing*, sponsored by US Army Materiel Command, *Summer Computer Simulation Conference*, San Francisco, Ca., 24 July 1975.

Spearman, F. R. J., "Analogue Computing Applied to Guided Weapons," *Discovery*, May 1957, pp. 192–197.

Spearman, F. R. J., J. J. Gait, A. V. Hemingway and R. W. Hynes, "TRIDAC, A large analogue computing machine", *Proc., IEE*, Vol. 103, Pt. B, 1956. p. 375.

Special Products News: The Beckman EASE Computer, Beckman Instruments Inc., South Pasadena, Ca., 18 April 1952.

Squires, M., "Interim Note on the Development of General Purpose Electronic Simulator", *RAE Technical Note: Guided Weapons*, No. 27, July 1948.

Staudenmaier, J. M., "What SHOT Hath Wrought and What SHOT Hath Not: Reflections on Twenty-five Years of the History of Technology", *Technology and Culture*, Vol. 25, no. 3, 1984, pp. 707–730.

Staudenmaier, J. M., *Technology's Storytellers: Reweaving the Human Fabric*, MIT Press, Cambridge, Mass., 1985.

Stein, M. L., "A General-Purpose Analogue–Digital Computer System", *Proc., Combined Analogue–Digital Computer Systems Symposium"*, Philadelphia, Pa., Dec 1960.

Stern, N., *From ENIAC to UNIVAC, An Appraisal of the Eckert-Mauchly Computers*, Digital Press, Bedford, Mass., 1981.

Stevenson, M. G., "Bell Labs: A Pioneer In Computing Technology: Part 1. Early Bell Labs Computers", *Bell Laboratories Record*, December 1973, pp. 345–350.

Stice, J. E., and B. S. Swanson, *Electronic Analog Computer Primer*, Blaisdell Publishing Co., New York, 1965.

Strang, C. R., *Computing Machines in Aircraft Engineering*, Douglas Aircraft Company, Santa Monica, Ca., 12 Dec 1951.

Stranges, A. N., "Farrington Daniels and the Wisconsin Process for Nitrogen Fixation", *Social Studies of Science*, Vol. 22, no. 2, May 1992, pp. 317–338.

Strong, J. D., "A Practical Approach to Analog Computers", *Instruments & Automation*, Vol 28, April 1955, pp. 602–610.

Terrett, D. S., "Makers of Analogue Computers," *Engineering*, 30 Aug 1963, p. 260.

The Lightning Empiricist, George A. Philbrick Researches, Boston, Mass., Vol. 1, no. 1, June 1952–Vol. 17, no. 3, Sept 1969.

The Physical Society, *Catalogue of the Thirty-Second Annual Exhibition of Scientific Instruments and Apparatus*, London, The Physical Society, 1948.

The Wall Street Journal, Monday, August 22, 1955, p. 15

Thomas, E. L., "A New Analogue Computor," *Engineering*, Vol. 176, Oct 1953, pp. 477–479.

Thomas, E. L., "Analogue Computation", *British Communications and Electronics*, May 1958, pp. 348–358.

Thomas, G. E., "Analogue Computers as Design Tools", *Automation Progress*, Vol. 3, Nov 1958, pp. 415–429.

Thomas, O. F., "Analogue–Digital Computers in Simulation with Humans and Hardware", *Proc., Western Joint Computer Conf.*, 1961, pp. 639–644.

Thomson, J., "On an Integrating Machine Having a New Kinematic Principle", *Proc., of the Royal Institute*, Vol. 24, 1876, pp. 262–265.

Thomson, W., "Harmonic Analyzer", *Proc., Royal Society*, Vol. 27, 1878, pp. 371–373.

Thomson, W., "Mechanical Integaration of the General Linear Differential Equation of Any Order with Variable Coefficients", *Proc., Royal Society*, Vol. 24, 1876, pp. 271–275.

Thomson, W., "Mechanical Integration of Linear Differential Equations of Second Order With Variable Coefficients", *Proc., of the Royal Institute*, Vol. 24, 1876, pp. 271–275.

Thomson, W., "On an Instrument for Calculating $\int \emptyset (x)\pi(x)\, dx$, the Integral of the Product of Two Functions", *Proc., Royal Society*, Vol. 24, 1876, pp. 266–268.

Tiechroew, D., and J. F. Lubin, "Discussion of Computer Simulation Techniques and a Comparison of Languages", *Simulation*, Oct 1967, pp. 181–190.

Todd, E. N., "Electric Ploughs in Wilhelmine Germany: Failure of an Agricultural System," *Social Studies of Science*, Vol. 22, no. 2, May 1992, pp. 263–282.

Tomayko, J. E., "Helmut Hoelzer's Fully Electronic Analogue Computer", *Annals of the History of Computing*, Vol. 7, no. 3, July 1985, pp. 227–240.

Tomovic, R., and W. J. Karplus, *High-speed Analog Computers*, John Wiley and Sons, New York, 1962.

Torrens, H. S., "A Study of 'Failure' with a 'Successful Innovation': Joseph Day and the Two-Stroke Internal-Combustion Engine", *Social Studies of Science*, Vol. 22, no. 2, May 1992, pp. 245–262.

Travis, I., "Differential Analyzer Eliminates Brain Fag", *Machine Design*, Vol. 7, No. 7, 1935, pp. 15–18.

Troe, J. L., "Nike in the air defence of our country", *Bell Laboratories Record*, Vol. 38, no. 4, 1960, pp. 122–129.

Truitt, T. D., "A Discussion of the EAI Approach to Hybrid Computation", *Simulation*, Oct 1965, pp. 248–257.

Truitt, T. D., "Hybrid Computation ... What is it? ... Who Needs it?...", *Spectrum*, Vol. 1, no. 6, June 1964, pp. 132–146; *Spring Joint Computer Conference*, 1964, pp. 249–269.

Turing, A. M., "On Computable Numbers, with an Application to the Entscheidungsproblem", *Proc., London Mathematical Society*, Vol. 42, 1936, pp. 230–265.

Tympas, A., "From Digital to Analogue and Back: The Ideology of Intelligent Machines in the History of the Electrical Analyzer, 1870s-1960s, *Annals of the History of Computing*, Vol. 18, no. 4, 1996, pp. 42–48.

Twigge, S. R., *The Early Development of Guided Weapons in the United Kingdom, Ph. D.* thesis, University of Manchester, 1990; published as Twigge, S. R., *The Early Development of Guided Weapons in the United Kingdom, 1940–1960*, Harwood Academic Publishers, London, 1992.

Van den Ende, J., *The Turn of the Tide: Computerization in Dutch Society 1900–1965*, Delft University Press, Delft, 1994.

Verzuh, F. M., "Memorandum on the MIT Differential Analyzer", *Typescript*, 10 Nov, 1954, MIT Archives AC62.

Vincenti, W. G., "Control-Volume Analysis: A Difference in Thinking Between Engineering and Physics," *Technology and Culture*, Vol. 23, no. 1, 1982, pp. 145–174

Vincenti, W. G., "The Air-Propeller Test of W. F. Durand and E. P. Lesley: A Case Study in Technological Methodology," *Technology and Culture*, Vol. 20, no. 4, 1979, pp. 712–751.

Vincenti, W. G., "The Davis Wing and the Problems of Airfoil Design: Uncertainty and Growth in Engineering Knowledge," *Technology and Culture*, Vol. 27, no. 4, 1986, pp. 717–758.

Vincenti, W. G., *What Engineers Know and How They Know It: Analytical Studies from Aeronautical History*, Johns Hopkins University Press, Baltimore, 1990.

von Braun, W., and F. I. Ordway III, *History of Rocketry & Space Travel*, Nelson, London, 1967.

von Neumann, J., *First Draft of a Report on the EDVAC*, Moore School of Electrical Engineering, University of Pennsylvania, 30 June 1945.

Wadel, L. B., and A. W. Wortham, "A Survey of Electronic Analog Computer Installations", *Trans., IRE, Electronic Computers*, Vol. EC-4, Pt. 2, June 1955, pp. 52–55.

Warshawsky, L. M., "Letter to the Editor", *The Simulation Council Newsletter*, The Simulation Council, Camarillo, Ca., March 1954, pp. 1–8.

Warshawsky, L. M., "My Career in Simulation", in J. McLeod (ed.), *Pioneers and Peers*, The Society of Computer Simulation International, San Diego, Ca., 1988, pp. 24–27.

Warshawsky, L. M., "WADC's New Large Analog Computer", *Proc., First Flight Simulation Symposium*, Nov 15–16, 1956, pp. 247–256.

Warshawsky, L. M., *The Simulation Council Newsletter*, in *Instruments & Automation*, Vol. 31, Nov 1958, pp. 1845–1850.

Warshawsky, L. M., *The Simulation Council Newsletter*, in *Instruments & Automation*, Vol. 32, Dec 1959, pp. 1861–1870.

Wass, C. A. A., *Introduction to Electronic Analogue Computers*, Pergamon Press, London, 1955.

Welbourne, D., *Analogue Computing Methods*, Pergamon Press, London, 1965.

Weyl, A. R., *Guided Missiles*, Temple Press, London, 1949.

White, M., *et al.*, "Trends in Simulation Hardware and Application: A Panel Discussion", *Simulation*, June 1971, pp. 279–285.

Wildes, K. L., and N. A. Lindgren, *A Century of Electrical Engineering and Computer Science at MIT, 1882–1982*, MIT Press, Cambridge, Mass., 1985.

Wilkes, M. V., "Automatic Computing", *Nature*, Vol. 166, 1950, pp. 942–943.

Wilkes, M. V., "The Coming of Multiple-Access Computers", *Journal of the Royal Aeronautical Society*, Vol. 71, no. 676, April 1967, pp. 235–236.

Wilkes, M. V., in D. R. Hartree, *Calculating Instruments and Machines*, The Charles Babbage Institute Reprint Series for the History of Computing, Vol. 6, Tomash Publishers, 1984.

Wilkes, M. V., *Memoirs of a Computer Pioneer*, MIT Press, Cambridge, Mass., 1985.

Williams, F. C., "Electro Servo Simulators," *IEE Journal*, Vol. 94, Pt. II A, 1947, pp. 112–129.

Williams, F. C., and A. M. Uttley, "The Velodyne", *IEE Journal*, Vol. 93, Pt. III A, 1946, pp. 1256–1274.

Williams, M., *A History of Computing Technology*, Prentice-Hall, Englewoods Cliffs, NJ., 1985.

Williams, R. W., "On the use of hybrid computing techniques in missile guidance computers", *Proc., Association International pour le Calcul Analogique, 4th Int., Conf.*, Brighton, Sept 1964, pp. 135–1 to 135–4.

Wilson, A. N., "Recent Experiments in Missile Flight Dynamics Simulation with the Convair 'ADDAVerter' System," *Proc., Combined Analogue–Digital Computer Systems Symposium*, Philadelphia, Pa, Dec 1960.

Wilson, A. N., "Use of Combined Analogue–Digital Systems for Re-entry Vehicle Flight Simulation", *Proc., Eastern Joint Computer Conference*, Washington DC, Dec 1961, pp. 105–113.

Winner, L., "Social Constructivism: Opening the Black Box and Finding it Empty", *Science as Culture*, Vol. 3, no. 16, pt. 3, 1993.

Winner, L., *Autonomous Technology*, MIT Press, Cambridge, Mass., 1977.

Wolin, L., "The Wolin–Saucier–Peak (WSP) Scientific Mix: A Quantitative Method for Comparing Hybrid and Digital Computer Performance", paper presented at, the *Special Symposium on Advanced Hybrid Computing*, sponsored by US Army Materiel Command, *Summer Computer Simulation Conference*, San Francisco, Ca., 24 July 1975.

York, H. *Race to Oblivion: A Participant's View of the Arms Race*, Simon and Schuster, New York, 1970.

Yoskowitz, L. K., "Methods of Pre-flight and Post-flight Simulation" *The Simulation Council Newsletter*, in *Instruments & Control Systems*, Vol. 34, April 1961, pp. 681–689.

Zagar, H., "Description of the Reeves Electronic Analog Computer (REAC)", *Mathematical Tables and Other Aids to Computation*, Vol. 3, no. 24, Oct 1948, pp. 326–327.

List of interviews

Professor Granino A. Korn, Tuscon, Arizona, 16 Oct 1990
Dr Walter J. Karplus, UCLA Computer Science Dept., Los Angeles, 9 Oct 1991
John McLeod, La Jolla, California, 11–12 Oct 1990
Professor Henry M. Paynter, Pittsford, Vermont, 27 Oct 1990
Dr George Barney, UMIST, Manchester, 15 March 1991
Jeff Baynham, Wellingborough, 30 April 1991
Derek J. Cartwright, Hendon, Sussex, 1 May 1991
Keith Knock, Manchester, 4 March 1991
Robin Penfold, BAe, Stevenage, 22 March 1991
Professor Howard Rosenbrock, UMIST, Manchester, 25 April 1991
Barry J. Tompson, BAe, Stevenage, 22 March 1991

Index

abacus 30
Aberdeen Proving Ground 108
AC amplifiers 64, 65, 73
AC network analyzers 19, 31
 (Table 2.1), 36–9 (Figure 2.1)
AC signal techniques 63
access and interaction 261–2
accuracy: classification scheme
 205–6; comparison between
 analogue and digital computers
 91, 182, 247–8, 257–61;
 comparison between
 mechanical and electronic
 systems 55–6; REAC
 computer 91; SIMLAC
 computer 205–6; speed and
 182; TRIDAC computer 182
ACE (Automatic Computing
 Engine) 51–2
Adams, W. G. 34
Adamson, A. L. 129
ADDAVERTER (Analogue-
 Digital-Digital-Analogue
 conVERTER) 149, 151
Admiralty Gunnery
 Establishment, Portland 185
Advanced Hybrid Computer
 Systems (AHCS) 170
Aeronautical Research Laboratory
 (ARL) 104–7
aeronautics: in Britain 21–2, 86,
 184–8; design 233–4; role of
 analogue computers 2, 56–7,
 63, 86, 134, 234; role of

computing 16–17; in USA
 22, 86
AGWAC (Australian Guided
 Weapons Analogue Computer)
 216n
Aitken, H. G. J. 276
Allen, D. W. 181
Ames Laboratory, Moffet Field
 128, 165, 166
amplifiers: AC 64, 65, 73; DC
 see DC operational amplifiers;
 number in USA by sector 136
 (Table 5.5); thermionic valve
 63, 64, 181, 271
analogue computers: access and
 interaction 237–8, 261–2;
 accuracy 55–6, 91, 182,
 205–6, 247–8, 257–61; all-
 analogue vs hybrid 263–4;
 applications (to 1960) 141
 (Table 5.7); centres established
 by American firms (1950–60)
 144 (Table 5.8);
 characterisation 30–3;
 commercial sales 91;
 comparison with digital
 computers 79–80, 91, 167–9,
 182, 213 (Table 6.5), 237,
 245–8, 266–7; complexity
 239–41; decline (1970s and
 beyond) 170–1; developers
 120–3; development 2, 29,
 271–3; developments (1960–70)
 151–64; developments

Printed in the United States
by Baker & Taylor Publisher Services

Printed in the United States
by Baker & Taylor Publisher Services